Energy, Climate and the Environment

The aim of this series is to provide texts which lay out the technical, environmental and political issues relating to proposed policies for responding to climate change. The focus is not primarily on the science of climate change, or on the technological detail, although there will be accounts of this, to aid assessment of the viability of various options. However, the main focus is the policy conflicts over which strategy to pursue. The series adopts a critical approach and attempts to identify flaws in emerging policies, propositions and assertions. In particular, it seeks to illuminate counter-intuitive assessments, conclusions and new perspectives. The intention is not simply to map the debates, but to explore their structure, their underlying assumptions and their limitations. The books in this series are incisive and authoritative sources of critical analysis and commentary, clearly indicating the divergent views that have emerged whilst also identifying the shortcomings of such views. The series does not simply provide an overview, but also offers policy prescriptions.

More information about this series at
http://www.palgrave.com/gp/series/14966

Rukonge Sospeter Muhongo

Energy Justice

A Local Content Analytical Framework for Sub-Saharan Africa

palgrave
macmillan

Rukonge Sospeter Muhongo
University of Dundee
Dundee, UK

Energy, Climate and the Environment
ISBN 978-3-030-61337-2 ISBN 978-3-030-61338-9 (eBook)
https://doi.org/10.1007/978-3-030-61338-9

Cover illustration: Antonio Cravo/EyeEm

This Palgrave Macmillan imprint is published by the registered company Springer Nature Switzerland AG
The registered company address is: Gewerbestrasse 11, 6330 Cham, Switzerland

ALL HONOUR AND GLORY TO GOD

Preface

The presence of oil and gas resources in Sub-Saharan Africa has brought both the gift and the curse. As Sub-Saharan African countries enjoy the abundance of natural resources such as oil and gas, on the one hand, these countries also experience distress. Oil and gas resources have proven to be vehicles of natural resources, changing deserts into tourist destinations, and the global north into an industrious technological powerhouse. But in Sub-Saharan African countries such as the DRC, Angola, Nigeria, Tanzania, Mozambique, Kenya, Liberia, Chad and Cameroon, these resources brought the paradox of plenty. They have created an indigenous capitalist elite cluster, with increased income inequality and poor economies. Oil and gas resources in Sub-Saharan Africa have not yielded similar results compared to countries like Norway and Brazil. Many factors have been mentioned, such as political instability, poor infrastructure, lack of technology, illiterate workforce and inability to service the oil and gas industry. Yet, the lucrative nature of the oil and gas industry could still salvage some development initiatives in Sub-Saharan Africa despite the circumstances. However, the local population normally lack access to the industry. The expensive and sophisticated nature of the oil and gas industry creates an enclave that the local community cannot pierce through without government intervention. This book makes an in-depth analysis of local content policies and their role in natural resource governance. Having Brazil and Norway as the comparative countries that have designed, adopted and implemented exemplary local content policies.

In Sub-Saharan Africa, most countries implemented a "one-size fits all approach" from the developed mature local content economies. The implementation of national content policy in Sub-Saharan Africa proliferated in the 1970s and 1980s up to date. The policy then concentrated on ownership of the resources without considering the nature of the global oil and gas market, the technological capability of the new indigenous owners, the absorptive capacity of the whole economy and the integration of the whole community. But because Sub-Saharan Africa had a strong nationalistic movement in the twentieth century, leading oil and gas producers like Angola and Nigeria adopted the "Indigenisation Policy of the 1970s". Mainly attributed to the nationalist wave of the 1970s and 80s, looking at history, one would say that local content policies are always compatible with the political environment and sentiment of the oil and gas host country. This book, argues that in the current wave of globalisation; low oil prices and the COVID-19 pandemic in addition to the current move away from fossil fuels to cleaner sources of energy, have disrupted the global and regional value chains. It is clear that Sub-Saharan Africa has an emergency of cracking through the oil and gas enclave and developing their economies. Unlike the 1970s and 1980s, Sub-Saharan Africa is more engaged in global and regional value chains, with massive inflows of foreign direct investments. To achieve and yield the most favourable outcome of these policies. Sub-Saharan Africa must adopt a regional content policy. Through a regional approach, Sub-Saharan Africa can leverage not only the natural resources but also the industrial parks, supplier clusters, regional financing mechanisms and regional training facilities. This will drive down the costs of production, increase efficiency and integrate the local Sub-Saharan population into the oil and gas industry.

Regional integration, in itself, is not enough to be a game-changer for Sub-Saharan African countries. Ranging from agricultural, economic, political and social reason Sub-Saharan Africa is proliferated with regional economic blocs. The regional content policy cannot be created independently of the regional economic blocs as this would raise issues of sovereignty, unifying, fiscal regimes as well as security concerns. It is within the East African Community, the Economic Community of West African Countries, the Common Market for East and Southern Africa, the South African Development Corporation and the African Union that such as policy can be created. Though a regional content theoretically provides a "new beginning" in local content adoption and implementation. The policy must be adopted within the realm of justice, by recognising the marginalised communities, ensuring provisions for equal distribution of ills and benefits of the oil and gas industry while following the right procedure.

Dundee, UK

Rukonge Sospeter Muhongo

Contents

Abbreviations

ABIMAQ	Brazilian Machinery and Equipment Association
ACFTA	African Continental Free Trade Area
AFDB	African Development Bank
BPD	Barrel per day
CNPE	National Energy Policy Council
COMESA	Common Market for Eastern and Southern Africa
COP21	United Nations Paris Conference
COVID-19	Corona Virus
CSOs	Civil Societies
CSR	Corporate Social Responsibility
DRC	Democratic Republic of Congo
EAC	East African Community
EACOP	East African Crude Oil Pipeline
ECOWAS	Economic Community of West African States
EIA	Environmental Impact Assessment
EITI	Extractive Industry Transparency Initiative
EPC	Engineering, Procurement and Construction
FDI	Foreign Direct Investment
FEED	Front End Engineering Design
GATT	General Agreement on Tariffs and Trade

GDP	Gross Domestic Product
GNP	Gross National Product
ICT	Information Communications Technology
IFC	International Finance Corporation
IGAD	Intergovernmental Authority on Development
IOCs	International Oil Companies
ISPs	Import Substitution Policies
LAPSSET	Lamu Port, South Sudan, Ethiopia Transport Corridor
LCPs	Local Content Policies
MEMD	Ministry of Energy and Mineral Development
MNCs	Multi-national Corporations
MSG	Multi-stakeholder group
NCS	North Continental Shelf
NEEC	National Economic Empowerment Committee
NNCP	Nigeria National Petroleum Corporation
NOC	National Oil Company
NORAD	Norwegian Agency for Development Cooperation
NYMBISM	Not in my backyard syndrome
OECD	Organisation for Economic Co-operation and Development
OFSE	Oil Field Service and Equipment
ONIP	Organização Nacional da Indústria do Petróleo
OPEC	Organization of the Petroleum Exporting Countries
PAU	Petroleum Authority of Uganda
PSA	Production sharing Agreement
PURA	Petroleum Upstream Regulatory Authority
R&D	Research and Development
SADC	Southern African Development Corporation
SEZ	Special Economic Zones
SLO	Social License to Operate
SMEs	Small Medium Entrepreneurs
WB	World Bank
WTO	World Trade Organisation

List of Figures

List of Tables

1

Local Content an Introduction

1.1 What Is Local Content?

Developing Sub-Saharan African countries have been the supplying hydrocarbons for over thirty years. Most of Sub-Saharan Africa has, incorporated standards for "local content" into the regulatory system regulating the production of natural resources. The goal is to create employment opportunities, facilitate the growth of businesses and acquire new skills and technologies.[1] Achieving such objectives weighs on the investor with increased costs and poses a threat for further investment. For example, oil and gas companies can face nationalistic demands that are unreasonable or deadlines from host governments or communities seeking fast outcomes; rewards come from long-term strategies focused on helping to grow local businesses and employees. Given the value of local content, several companies consider the idea a strategic one that can directly influence a variety of core business functions,

[1]Ablo, Austin Dziwornu, 2015. Local Content and Participation in Ghana's Oil and Gas Industry: Can Enterprise Development Make a Sifference? In *The Extractive Industries and Society*.

R. S. Muhongo, *Energy Justice*, Energy, Climate and the Environment,
https://doi.org/10.1007/978-3-030-61338-9_1

including market growth, acquisition and operations.[2] Besides, investors have begun developing creative technologies and methods to meet the requirements of local content and contribute to sustainable economic development.

> Local content is the added value brought to a host nation (and regional and local areas in that country) through the activities of the oil and gas industry. This may be measured (by project, affiliate, and/or country aggregate) and undertaken through workforce and local supplier development.[3]

Local content emphasises on value addition, or what is known as in country value addition that develops other sectors of the economy. The competing interests of regional and/or local areas that are host communities of oil and gas resources have brought about major debates. That we clarify using justice as an analytical framework for local content implementation. The fundamental principle behind adopting a local content policy is cracking through the enclave of the oil and gas sector by developing the local population. Little is being mentioned on what metric is most suitable for developing, growing nascent oil and gas producers, such as Liberia, Tanzania, Kenya, Uganda and Mozambique. Local content has been pivotal in transforming oil and gas resources into drivers of economic prosperity.

1.2 Why Is Local Content Important?

Regardless of boom and bust cycles, market fluctuations, uncertain pandemic outbreaks such as COVID-19 and other factors, the many new oil and gas producers in Africa will take advantage of these policy's

[2] Ibid.

[3] UNDP, 2012. *UNDP's Strategy for Supporting Sustainable and Equitable Management of the Extractive Sector for Human Development*. New York.

potential benefit and relevance for strategies to offer resource-rich developed countries petro-development.[4] While local content policies are only one component of a more holistic natural resource-based development system, they are extremely significant. For local content to thrive where previous efforts of resource nationalism have failed, actors writing, advising, controlling and implementing these policies must learn from the shortcomings of those earlier policies. Both modern and old local content strategies need to be updated to resolve crucial issues surrounding the meaning and evaluation of local content.[5] While those creating newer local content policies need to continue to pursue a balance between strict regulation and promoting investment to ensure the best possible outcome.[6] Financial capital and local leaders must also align their interests with the needs of ordinary people as well as societies that host oil and gas operations. Host governments and international oil companies must connect local content to governance, anti-corruption policies and meaningful transparency by all parties—while donor agencies and international institutions may assist them. Local content policies can be transformative, with the right synergy with other policies and participation of all stakeholders. Development policies-based solely on oil rents have, in every case disappointed Sub-Saharan Africa. However, local content provides new opportunities to make oil and gas work for twenty-first-century African production.

1.3 Channels of Local Content Policies

Local content policies are found in policy statements, legislation, oil and gas contracts and legislation. But local content policies can only be implemented through three channels, namely national, regional and local

[4]Ovadia, Jesse Salah, 2012. The Dual Nature of Local Content in Angola's Oil and Gas Industry: Development vs. Elite Accumulation. *Journal of Contemporary African Studies*, 30(3), pp. 395–417.

[5]Ovadia, J.S., 2015. The Role of Local Content Policies in Natural Resource-Based Development. *Rohstoffe und Entwicklung*, 37.

[6]Kolstad, I. and Kinyondo, A., 2017. Alternatives to Local Content Requirements in Resource-Rich Countries. *Oxford Development Studies*, 45(4), pp. 409–423.

channels of implementation, in Sub-Saharan Africa leading local content regimes in Nigeria and Angola.[7] Have adopted the national content policy; of local content implementation. New oil and gas producers like Mozambique, Tanzania, Kenya, Uganda and Liberia have also followed suit. Without an in-depth analysis of what would work for their circumstances. We advocate for the implementation of the regional content policy. The regional channel of implementation seems more viable for nascent oil and gas producers. Currently, the global oil and gas industry has faced the ramifications of the COVID-19 pandemic and the oil price shocks of 2020. These challenges have disrupted global value chains and investments in the oil and gas industry. Several countries have signed the Paris Agreement, gearing the wheels of the energy transition. Global consumers of fossil fuels moving to cleaner sources of energy with strict time-lines of 2030, 2040, 2050, etc. Sub-Saharan new oil and gas new participants do not have the luxury of time to develop the skills, capital, absorptive capacity, and technology transfer, as well as an adequate supplier base to crack through the enclave and integrate the oil and gas industry. Through regional integration countries in the Gulf of Guinea, in East Africa, in Southern Africa can pool resources, create clusters, develop regional training facilities and provide regional financing mechanisms that can facilitate local integration into the oil and gas value chain with massive linkages to other regional sectors of the economy.

1.4 Purpose of the Book

Since the African nationalisation wave of the 1980s, Sub-Saharan Africa has struggled to transform our vast natural resources into vehicles of economic development. Several studies have been undertaken explaining the paradox of plenty, trying to assess as to why Sub-Saharan Africa a rich region, with a growing young population is still deep into poverty. Several countries adopted the "one-size fits all" approach addressing this challenge in Sub-Saharan Africa. Without considering the pre-existing

[7] Nwapi, C., 2016. A Survey of the Literature on Local Content Policies in the Oil and Gas Industry in East Africa. SPP Research Paper, (9/16).

circumstances to make these "solutions" and policy options thrive in our continent. This book offers novel ideas as to how policies such as the local content policy can be adopted and implemented in Sub-Saharan Africa and avail real solutions to local integration in the oil and gas industry. The use of regional integration as a problem-solving tool has been used in Sub-Saharan Africa in issues of politics, economics, and trade but not specifically in the oil and gas industry. Making it easy for multinational oil and gas companies have higher bargaining power in Sub-Saharan Africa. The oil and gas industry in Sub-Saharan Africa has brought about corruption, income inequality and built an indigenous capitalist elite class. The growing inequality, lack of transparency and negative impact of the oil and gas industry in Sub-Saharan Africa motivate the research done in this book to merge local content policies and energy justice. Having the right procedures, recognising marginalised communities, ensuring equitable distribution of the benefits and ills of the oil and gas operations with the right time and location.

1.5 Methodology

This book analyzes the adoption and implementation of local content regimes in Sub-Saharan African oil and gas countries. It is essentially an examination of how developing nations and developed nations have approached, or are likely to approach the process under which local content policies have been implemented. This study is doctrinal research (with certain non-doctrinal elements) that utilises the comparative case study methodology. Kaarbo and Beasley define comparative analysis as "the systematic comparison of two or more data points ('cases') obtained through use of the case study method".[8] They also confirm that case studies can be both qualitative as well as narrative, and do not necessarily need to rely on multiple sources of evidence to function. The functionalism in comparative law places emphasis not only on legal

[8]Kaarbo, J. and Beasley, R.K., 1999. A Practical Guide to the Comparative Case Study Method in Political Psychology. *Political Psychology*, 20(2), pp. 369–391.

systems and rules but as well as the effects of the legal systems.[9]. The function of local content policy will be evaluated based on their functionality in the local host community of the energy resources. This book aims to unify different legal frameworks and create a new policy that can be applied in any oil and gas region, especially in regions that are struggling to crack through the oil and gas industry enclave. The analysis will examine whether the new preferable policy, institution, theory or result could work equally well in another oil and gas regions given the same circumstances. The selection process of the countries was based on the similarities of the countries. The similarities between the Sub-Saharan African countries such as Liberia, Kenya, Angola, Equatorial Guinea, Nigeria, Mozambique, Tanzania and Uganda are the following: all members of the same international agreements, namely the East African Treaty, the Economic Community of West African States, the African Continental Free Trade Area, The General Agreement on Tariff and Trade (GATT), etc. These countries have all adopted and implemented the same typology of local content policy, namely the national content policy, with similar political background and are semi-democratic countries. The enquiry and comparative method employed will examine the micro and macro comparison aspects in adopting local content policies.[10] Hence the unification of local content requirements between the Sub-Saharan African countries will look at the economic, social, political and cultural factors in which legal rules are embedded beyond other societal preferences when implementing such production development policies.

The analysis done under a comparative enquiry is useful for observing the gaps in the adoption and implementation process of local content policies. The gap can be found not only in the adoption and the implementation process. But also in the fact that the adopted policies are not just. The analysis requires the countries to borrow a leaf from the local content poster boys on developing exemplary local content policies. But such transplantation must be done in line with the prevailing

---

[9]Dubber, M., Reimann, M. and Zimmerman, R., 2006. *The Oxford Handbook of Comparative Law*.

[10]Dubber, M., Reimann, M. and Zimmerman, R., 2006. *The Oxford Handbook of Comparative Law*.

circumstances of each country's context. For instance, specific production development policies may thrive in stable political conditions, strong, self-reliant economies, technological sound environment, etc. These factors may not be readily availed in other Sub-Saharan countries. Calling for novel policy suggestions that meet the demands of the oil and gas industry and the local population. No literature creates a comparison analysis of local content policies for the sole purpose of creating a regional content policy, especially between Norway, Brazil and Sub-Saharan Africa.

1.6 Book Overview

This chapter—gives a brief overview of what the general understanding of local content is, the channels of local content implementation. The chapter also provides the methodology used and the purpose of writing this book.

Chapter 2—explains the proliferation of local content policies in numerous oil and gas countries, as a means of government intervention into the oil and gas industry to increase domestic participation. Moreover, local content policies have traditionally been seen as a tool that transforms natural resources such as oil and gas into an industry that develops the whole (national) economy. The chapter postulates that local content policies yield different results in developing countries like Nigeria, Kenya, Uganda and Tanzania, unlike in developed countries like Norway and Brazil. Using a case study approach, we analyse local content policy tools and how these tools affect local content design during the implementation of these policies. This chapter also introduces energy justice as a theory that can guide the adoption and implementation of local content policies. The purpose of having energy justice in the adoption of these policies is to shift away from the traditional political and economic justification of having local content policies and have justice as a pivotal motivator of adopting local content policies.

Chapter 3—expounds on the theory of "Energy justice" which has been a pivotal aspect in energy transitions. In this chapter, energy justice is incorporated into the adaptation of local content policies in the oil and gas countries. The principles of energy justice, namely justice as recognition, distributive justice and procedural justice, are addressed to mediate the harm done by oil and gas operations, especially in developing oil and gas-rich countries such as Nigeria, Liberia, Equatorial Guinea, Angola, Kenya, Uganda, Mozambique and Tanzania. The main contribution is analysing whether Norway and Brazil have elements of justice in their policies and whether this is among the main factors of the policy's success as compared to local content policies adopted in other regions. The principles of energy justice are restorative, and they enhance restorative justice in policy formulation to resolve the harm suffered by the domestic stakeholders. This poses questions about why these policies yield different results in different countries. The chapter expounds on the methodological development of the book as well as the application of local content policies and energy justice within the different tiers of energy law driving the research towards the objectives of the book.

Chapter 4—Traverses different theoretical constructs in trying to understand local content policies. These constructs are based on a localism or a regional approach advocating for either regional integration mechanisms or local mechanisms of implementing local content policies. This is guided by the understanding that numerous factors affect local content policy success both in developing and developed oil and gas countries. Nevertheless, local content as an interdisciplinary phenomenon navigates through different economic and legal schools of thought. The interdisciplinary nature of these policies creates uncertainty on what type of local content policy should best address the needs of the oil and gas industry in a given jurisdiction. National content policies have been fundamentally the main type of local content policies implemented across the book's case study countries. This has led to different alternatives being adopted in oil and gas countries that move away from local content policies to address the industry bottlenecks and achieve the same objectives as local content policies. A robust literature review is done in this chapter to find the gap both in literature and policy, to meet the objectives of the book and create novel solutions for developing countries when adopting local content policies.

Chapter 5—undertakes a cross-country study wherein analysis is made to determine how local content policies function in different jurisdictions: all case study countries, namely Nigeria, Tanzania, Norway, Brazil, Kenya, Angola, Mozambique, Equatorial Guinea, Liberia and Uganda have all adopted national content policies. But only the national content policies in Norway and Brazil are efficient and recommendable to other economies. This brings about the question as to whether these policies are only efficient in developed nations compared to developing nations as they yield different results. As such, the chapter provides a cross-country analysis for local content policies. Furthermore, the chapter analyses the initial government intervention mechanisms adopted by all countries, the economy before the discovery of oil and gas resources and the local content regime found in all six countries. The principles of energy justice are assessed in all the local content regimes to determine whether principles of energy justice play a role in making local content policies effective and efficient in both developing and developed countries.

Chapter 6—explains the concept of regionalism in a local content perspective and its rationale and role in achieving the development objectives of mostly the developing countries in Sub-Saharan Africa. Oil and gas countries such as Norway, Brazil and Nigeria have implemented national content policies since the discovery of their oil and gas resources. But these policies have not brought about similar results despite the similarity in implementation and adoption mechanisms of the policy, since developing countries like Kenya, Tanzania and Uganda do not have the same characteristics of developed countries like Norway and Brazil. For these nascent oil and gas countries to avoid the Nigerian route of implementing redundant local content policies, it is argued here that these countries should adopt a regional content policy. Regional content policies pool resources of the countries within a given regional economic bloc to meet the demands of the oil and gas industry and have leverage against the huge multinational oil and gas corporations—the principles of energy justice guide regional content policies. The chapter also hypothesises that as much as policymakers can implement local content policies based on economic and political reasons, the aspect of justice must be at the core centre of adopting these policies.

Chapter 7—Examines the implications of regional integration on local content policies. The chapter further illustrates how individual countries can gain from having a regional economic bloc as the centre for local content development as the chapter shows how the policy can be developed in the East African Community. The key contribution of this chapter is the analysis done of regional content policies and stakeholder engagement from the international level to the community level. Secondly, the chapter explains the role between regional content policies in achieving sustainable development goals and the implications of having energy justice within a local content framework are all illustrated in this chapter. A regional content policy is a typology of local content policy. Regional content policies are yet to be implemented in any oil and gas-rich region. The policy entails the pooling of resources by the indigenous oil and gas stakeholders of a particular regional bloc such as the East African Community to compete with the oil and gas multinational investors. The policy encompasses all stakeholders of the oil and gas industry.

Moreover, stakeholders emanate from the regional level, national level and the immediate community level. As such, the chapter construes that it is only a regional content policy that can have an impact on all levels of stakeholder engagement in any local content policy. This proves that a regional content policy has ramifications that are in line with the demands of energy justice, in that the policy recognises all stakeholders through a justifiable process that brings about an equal distribution of ills and benefits of the oil and gas industry. Furthermore, the chapter also analyses the implications of the policy on the sustainable development goals, the role of the countries in the region, the regional value chains as well as the linkage development in East Africa's oil and gas sector.

Chapter 8—Local content policies have been identified to be weak in developing Sub-Saharan African compared to Brazil and Norway. The weakness is evident from the first stages of implementation in these developing countries, the misalignment of policies, and the heavy involvement of the political elite in oil and gas operations. The book provides ample lessons for Sub-Saharan African oil and gas countries from mature developed oil and gas economies, namely Brazil and Angola. The chapter also provides an insight into regional content policies and uncertain global events such as the 2020 oil price shock and the COVID-19 global pandemic that disrupted global and regional value chains.

References

Ablo, Austin Dziwornu, 2015. Local Content and Participation in Ghana's Oil and Gas Industry: Can Enterprise Development Make a Difference? In *The Extractive Industries and Society*.

Dubber, M., Reimann, M. and Zimmerman, R., 2006. *The Oxford Handbook of Comparative Law*.

Kaarbo, J. and Beasley, R.K., 1999. A Practical Guide to the Comparative Case Study Method in Political Psychology. *Political Psychology*, 20(2), pp. 369–391.

Kolstad, I. and Kinyondo, A., 2017. Alternatives to Local Content Requirements in Resource-Rich Countries. *Oxford Development Studies*, 45(4), pp. 409–423.

Nwapi, C., 2016. A Survey of the Literature on Local Content Policies in the Oil and Gas Industry in East Africa. SPP Research Paper (9/16).

Ovadia, Jesse Salah, 2012. The Dual Nature of Local Content in Angola's Oil and Gas Industry: Development vs Elite Accumulation. *Journal of Contemporary African Studies*, 30(3), pp. 395–417.

Ovadia, Jesse Salah, 2015. The Role of Local Content Policies in Natural Resource-Based Development. *Rohstoffe und Entwicklung*, 37.

UNDP, 2012. *UNDP's Strategy for Supporting Sustainable and Equitable Management of the Extractive Sector for Human Development*. New York.

2

The Different Aspect of Local Content

2.1 Introduction

Local content policies entail that international firms producing any goods and services within a country are to acquire a given quantity or percentage of intermediate inputs domestically.[1] These policies aim to integrate the local population into the energy value chain and maximise the probable advantages of foreign direct investment. They are protectionist, similar to import quotas and tariffs on imported goods.[2] Local content policies advocate for the participation of local goods and services, local employment and use of local raw materials to retain value addition by the foreign investors and other market participants in the extractive industry value chain. The presence of oil and gas resources is perceived as an opportunity to boost a country's economic advancement, and local content policies can play a pivotal role in making this a reality. These

[1] Nwapi, C., 2015. Defining the "Local" in Local Content Requirements in the Oil and Gas and Mining Sectors in Developing Countries. *Law and Development Review*, 8(1), pp. 187–216.
[2] Ibid.

R. S. Muhongo, *Energy Justice*, Energy, Climate and the Environment,
https://doi.org/10.1007/978-3-030-61338-9_2

policies can also be considered as production development policies.[3] For local content policies to thrive within a given economy, they need to be compatible with other policies such as procurement and industrial policies. The existence of local content policies has brought about economic development to economies that have a stable democracy, the rule of law, absorptive capacity, technological awareness, as well as a credible and stable economic climate. Most oil and gas are not found in such economies with such favourable characteristics for local content policies to thrive. Countries such as Norway and Brazil have adopted and implemented among the best local content policies in the world due to their supervening characteristics before finding the oil and gas. In contrast in Sub-Saharan African economies such as Kenya, Tanzania, Angola, Mozambique, Nigeria and Uganda should be able to adopt local content policies in an unconventional manner that is suitable and tailored to their local needs.

This chapter covers the history of local content policies, the policy tools that affect the design of local content policies will be covered in this chapter. Furthermore, the chapter will also cover the level of government intervention needed when implementing local content policies, the significant benefits and disadvantages of local content policies according to the literature reviewed.

2.2 The History of Local Content Policies

By the end of World War Two, most countries were emerging from economic turmoil. They established productive development policies, which were also known as import substitution policies (ISPs).[4] ISPs encouraged the reallocation of resources from traditional industries onto the advanced industries.[5] This facilitated higher productivity in the economy associated with clustering (also known as agglomeration

[3]Monge-González, R., Rivera, L. and Rosales-Tijerino, J., 2010. Productive Development Policies in Costa Rica: Market Failures, Government Failures, and Policy Outcomes.

[4]Rodriguez-Clare, A., 2007. Clusters and Comparative Advantage: Implications for Industrial Policy. *Journal of Development Economics*, 82(1), pp. 43–57.

[5]Ibid.

economies) in the given industry. ISPs were regarded as inefficient and placed difficulties in industries being competitive. ISPs became unpopular due to two main reasons. According to Rodriguez-Claire "advanced economies have already reaped the benefits of clustering in the advanced sector".[6] This lowers the prices of goods in the international market, due to high productivity in advanced economies associated with clustering of economies. Therefore exported goods from developing countries have no competitive advantage in the international market. The second reason is that ISPs lead to assumptions that the production made in advanced industries always leads to clustering.[7]

Lall clarifies the unpopularity of ISPs by stating that "the protection mechanism is not effective because ISPs are not aligned to other policies.[8] To increase domestic competition and thereby avoid complacency among protected companies".[9] ISPs have primarily been abandoned, but most countries do maintain some characteristics similar to ISPs such as fiscal regimes being lenient on the exports of goods and services, foreign direct investments, tax incentives towards specific industries. This led to a new wave of production development policies known as local content policies. These policies were first implemented in the UK North Sea region in the 1970s.[10] Local content policies apply to both the international oil companies (IOCs) and the local companies. They can be implemented to restrict imports as well as create national oil companies. Local content policies also advocate for the creation of both backward and forward linkages. These policies have recently been used to facilitate economic diversification.[11]

[6]Rodriguez-Clare, A., 2007. Clusters and Comparative Advantage: Implications for Industrial Policy. *Journal of Development Economics*, 82(1), pp. 43–57.

[7]Hausmann, B.M., Moreira, M. and Rodrik, D., 2006. Coordination Failures, Clusters, and Microeconomic Interventions. *Andrés Velasco*, p. 1.

[8]Lall, S., 2013. Reinventing Industrial Strategy: The Role of Government Policy in Building Industrial Competitiveness. *Annals of Economics & Finance*, 14(2).

[9]Ibid.

[10]Heum, P., 2008. Local Content Development: Experiences from Oil and Gas Activities in Norway.

[11]Tordo, Silvana, Warner, Michael, Manzano, Osmel E. and Anouti, Yahya, 2013. *Local Content in the Oil and Gas Sector (English). A World Bank Study*. Washington, DC: World Bank.

China's local content policies are regarded as successful in diversifying the Chinese economy. Rodrick argues that "the Chinese government emphasised policy interventions have facilitated the development of domestic capabilities in consumer electronics and other technologically advanced industries".[12] Hence local content policies help an economy look beyond a single industry. Local content policies are adopted and implemented due to numerous factors. Rent-seeking factions of a given country can adopt them. These factions adopt local content policies to "legalise" their conducts. Ovadia states that "the Angolanization policies of the oil and gas industry in Angola served multiple purposes giving the political elite the upper hand in influencing government policy for their benefit".[13]

For example, in Venezuela, a law was passed that forced all the international oil companies to refine oil in Venezuela.[14] While in Brazil President Vargas in 1953 declared Petrobras must use local capital, local people and local know-how.[15] Heum states that "in essence, the key to Norway's industrial success has not adopted from a protectionist market, but to arrange for a dynamic industrial and technological development that involves competent actors within the domestic knowledge base and leading international competence".[16] Local content policies have frequently been argued to be against international trade measures that apply to the member states for the World Trade Organization (WTO).[17] Particularly the agreement on Trade-Related Trade Organization, the General Agreement on Tariff and Trade (GATT), and the Agreement on Subsidies and Countervailing Measures (ASCM). This

---

[12] Rodrik, D., 2006. What's So Special About China's Exports? *China & World Economy*, 14(5), pp. 1–19.

[13] Ovadia, J.S., 2012. The Dual Nature of Local Content in Angola's Oil and Gas Industry: Development vs. Elite Accumulation. *Journal of Contemporary African Studies*, 30(3), pp. 395–417.

[14] Tordo, Silvana, Warner, Michael, Manzano, Osmel E., and Anouti, Yahya, 2013. *Local Content in the Oil and Gas Sector (English). A World Bank Study*. Washington, DC: World Bank.

[15] Ibid.

[16] Heum, P., 2008. Local Content Development: Experiences from Oil and Gas Activities in Norway.

[17] Hufbauer, G.C., Schott, J.J. and Cimino-Isaacs, C., 2013. *Local Content Requirements: A Global Problem* (Vol. 102). Columbia University Press.

principle holds that a state should treat another state as they would treat their nationals.[18] These agreements are applied differently to different nations since not all countries have the same economic development. They create exceptions for least developed and developing nations.[19] The Agreement on Subsidies and Countervailing explicitly restrict the use of subsidies, if only they "adversely" cause adverse effects to another WTO member in their application.[20] It is difficult for party bearing the onus of proof to prove how the subsidies "adversely" affect their domestic economy. This is because what is regarded as "adversely" affected in this agreement is narrow. Its concluded that due to the low economic development in developing and least developed countries, it would only be fair to create a "policy space" that allowed these countries to reach their economic goals. The principle of "reasonable accommodation" has allowed the use of these policies.[21] This has made WTO rules regulating the use of local content policies not be stringent and strict in their application.

2.3 Local Content Policy Tools

As it has been mentioned earlier that local content policies just like agricultural policies, industrial policies and procurement policies are driven political policies. Government adopts and designs the local content policy according to what the government's needs or "deems fit". Local content policies are found in a country's legal framework. The policy can be embedded in Acts of Parliament, and other provisions such as policy, regulations and contracts. For example, Article 50 of the Law on Subsoil Users in Kazakhstan states that "bids for new subsurface mineral rights include commitments to minimum levels of local content

[18]Ibid.

[19]Ramdoo, I., 2015. Unpacking Local Content Requirements in the Extractive Sector: What Implications for the Global Trade and Investment Frameworks. By International Centre for Trade and Sustainable Development (ICTSD).

[20]Ibid.

[21]Ramdoo, I., 2015. Unpacking Local Content Requirements in the Extractive Sector: What Implications for the Global Trade and Investment Frameworks. By International Centre for Trade and Sustainable Development (ICTSD).

in goods, works and services, and the training of Kazakh personnel".[22] Local content can also be found in secondary legislation, for example, in Angola. Ministerial Order No. 127/03 on the Contracting of local Services and local Goods from local Companies by international oil companies.[23] These policies can also be in contractual obligations such as the 2008 model production sharing agreement (PSA) that was adopted in Tanzania sets local content requirements as part of the contractual obligations.[24] The state can use various policy tools in implementing local content policies.

These tools include the following[25]:

I. Contractual obligations that call for the inclusion of local goods and services
II. Contractual obligations that call for training of the local populace
III. Regulatory or contractual measures that foster the technological transfer
IV. The different bidding parameters that call for local content provisions as a prerequisite for being awarded a license
V. Attracting reinvestment of the profits from international oil companies into the local economy
VI. Regulations and contractual obligations on infrastructural and educational provisions
VII. Provisions for local ownership
VIII. The direct governmental intervention through state-owned enterprises (SOEs).

[22] Law on Subsoil and Subsoil Users in Kazakhstan, No. 291-IV, June 2010, Republic of Kazakhstan.

[23] General Regulatory Framework for Hiring of Services and Goods from National Companies in the Oil Industry ("Decree 127/03").

[24] http://www.tpdc-tz.com/MPSA%20_2008.pdf Art. 18 of the PSA that calls for maximization of local goods and services where available on a competitive basis.

[25] Tordo, Silvana, Warner, Michael, Manzano, Osmel E. and Anouti, Yahya, 2013. *Local Content in the Oil and Gas Sector (English). A World Bank Study*. Washington, DC: World Bank.

2.3.1 Policy Tools Affecting Local Content Design

Policy tools determine the governance of local content policies. In the hierarchy of the legal framework, the policy tools may be implemented based on the contractual requirements, depletion policies, industrial policy, procurement policy, tax incentives and local industry participation capabilities.[26] These policies are implemented to facilitate value creation in the local industry. But these policies can also have inefficiencies since they are purely politically driven. This can lead to corruption.[27] Local content policies can also be ineffective due to "technological strangeness".[28] This is why the two local content forefront Sub-Saharan countries, namely Nigeria and Angola, implemented local content policies with little success. Countries like Nigeria and Angola do not possess the ability to meet the sophisticated necessities of the oil and gas industry. This makes it challenging for such countries to develop forward and horizontal linkages in their local economy.[29] Hence the minimal value is being added in the local economy, due to such shortfalls. It is clear that to adopt and implement local content policies; there is a need for government intervention. Since developing countries do not have the market structure that automatically attracts local sourcing of goods and services as well as local participation at a competitive price, timeous delivery and adequate quality; there is a need for governmental intervention.

Tordo states that "governments must decide on how to explore the resources while maximising social welfare according to the resources available".[30] For example, Norway chose to develop its oil and gas resources slowly compared to the United Kingdom. The United Kingdom developed their resources faster than Norway. This was because

[26]Tordo, Silvana, Warner, Michael, Manzano, Osmel E., and Anouti, Yahya, 2013. *Local Content in the Oil and Gas Sector (English). A World Bank Study*. Washington, DC: World Bank.

[27]Campos, J.E. and Pradhan, S. (Eds.), 2007. *The Many Faces of Corruption: Tracking Vulnerabilities at the Sector Level*. World Bank Publications.

[28]Technological strangeness is defined as the ability of the rest of the economy to develop service capacity through backward linkages and the speed at which such capacity can be created.

[29]Ramdoo, I., 2013. Fixing Broken Links.

[30]Tordo, Silvana, Warner, Michael, Manzano, Osmel E. and Anouti, Yahya, 2013. *Local Content in the Oil and Gas Sector (English). A World Bank Study*. Washington, DC: World Bank.

Norway decided to develop its service industry slowly.[31] This facilitated local inclusion, and as a result, the local service firms became internationally competitive. While the United Kingdom chose to attract American service companies and expertise.[32] In Sub-Saharan economies, governments do not have the luxury to slowly develop their resources, since most of these countries are developing countries. They need funds in their government coffers to provide social services to their population. The design of local content policies is fundamentally based on the design of their depletion policies and how these principles are intertwined with the local content design within any given economy.

2.4 Impact of Depletion Policies on Local Content Design[33]

2.4.1 Introduction

Depletion policies are policies that regulate the pace at which the oil and gas resources will be explored, produced and developed. These policies affect the whole design of local content policies since a country can choose what type of policies to implement whether the policies should be hard strict policies based on steep qualitative targets or soft value-based local content policies.

I. Good oilfield practise: reservoirs may be damaged if the oil fields are not properly maintained.
II. Politics: The country may enter international agreements that affect productive capacity and output. Hence affecting the decision making with regard to a government's resources. This has been

[31] Al-Kasim, F., 2006. *Managing Petroleum Resources: The "Norwegian Model" in A Broad Perspective* (Vol. 30). Oxford Institute for Energy Studies.

[32] Al-Kasim, F., 2006. *Managing Petroleum Resources: The "Norwegian Model" in a Broad Perspective* (Vol. 30). Oxford Institute for Energy Studies.

[33] Stiglitz, J.E. and Dasgupta, P., 1981. Resource Depletion Under Technological Uncertainty.

evident in the current OPEC and Russia agreed to cut down oil production.

III. State budget: The need for public finances may speed up the production schedule.

IV. Political pressure on public spending: an increment in public income can lead to increased public spending, without regarding the appropriate reinvestment opportunities.

V. Domestic economy: The domestic economy must have the absorptive capacity for reinvestment opportunities. Otherwise, this might lead to hyper-inflation and rapid changes in foreign exchange rates (Dutch disease). The domestic economy may also not be able to foster linkages to the local economy.

VI. Institutional structure and national governance: the unavailability of transparency in the institutional structure, proper checks and balances leads to the inappropriate implementation of policies. In jurisdictions with low transparency, weak bureaucracy it is better if the resources were left in the ground.

VII. Resource curse: this links both the local economy and the institutional structure of a country. This is the failure to propel the oil and gas resources into a driver of sustainable economic development.

VIII. Price expectations: the volatility of resource prices affect the speed at which a government may choose to develop their oil and gas resources.

IX. Cost expectations: when costs of developing oil and gas resources are high. Most governments tend to wait for technological advancements that will decrease the costs.

X. Time value of money: Victor states that "the earlier the extraction date, the higher the potential gains, ceteris paribus".[34] The un-proven reserves do not automatically gain an income. This is dictated by the potential return in the investments thereof.

[34]Victor, N.M., 2007. On Measuring the Performance of National Oil Companies (NOCs). Program on Energy and Sustainable Development working papers, WP64, Stanford University.

> The factors mentioned above affect the way a government adopts local content policies. In the 1950s, the service and supply industries of Brazil could not meet the demands of the oil and gas industry.[35] The government required North American services and supply companies to set up operations in Brazil. Nordas state that "most of the designs and equipment used in refining or petrochemical processing, pipelines and oil field production came from these North American companies".[36] This helped the adopted local content policy foster forward and horizontal linkages quickly rather than concentrating only on backward linkages. The presence of political and ethnic civil unrest in Indonesia decreased foreign direct investments.[37] This led to minimal linkages, minimal spillovers, and minimal transfer of technology. The result was that Indonesia adopted an untraditional local content policy. The policy only concentrated on the fact that the production should take place in the domestic territory, regardless of whether the company was controlled and owned by foreign investors or not.

The state-budget, pressure for public spending and the domestic economy also affect the design of local content policies. Nigeria in the oil boom period had an increase in foreign exchange earnings.[38] The oil proceeds financed the steel mills in the 1970s this accelerated government revenue brought about large infrastructural projects.[39] The Nigerian economy did not have the absorptive capacity to contain the revenue streams and benefits from the oil and gas industry. The Nigerian government created steel mills in excess capacity compared to the local demand. The policy concentrated on local ownership of oil and gas reserves rather than developing local capabilities that could

---

[35] Mendonça, R.W. and de Oliveira, L.G., 2013. Local Content Policy in the Brazilian Oil and Gas Sectoral System of Innovation. *Latin American Business Review*, 14(3–4), pp. 271–287.

[36] Nordås, H.K., Vatne, E. and Heum, P., 2003. The Upstream Petroleum Industry and Local Industrial Development: A Comparative Study.

[37] Tordo, S., 2011. *National Oil Companies and Value Creation.* The World Bank.

[38] Gidado, M.M., 1999. Petroleum Development Contracts with Multinational Oil Firms: The Nigerian Experience. Ed-Linform Services.

[39] Ibid.

create sustainable development in Nigeria. The capability of the local market participants to be involved in the oil and gas industry is essential, especially for the domestic economy. The local market participants are either state-owned entities that have no external participation as well as competing market participants with no governmental intervention. Local content policy requirements are met differently by the two different types of entities. Hansen states that "in the 1990s and 2000s African economies had minimal interventions in the oil and gas industry. This attracted FDI, privatisation and the development of a stable, transparent business-friendly environment".[40] The design and consideration of such government intervention are vital in this Covid-19, oil price shock era of 2020. COVID-19 has challenged extractive supply chains model for the sourcing of goods and labour. The current health crisis has brought trade and industry to an abrupt halt. It forces countries to develop novel and unconventional solutions on how to safeguard the supply of goods and services, both of which depend on a healthy workforce. Resource-rich countries need to start developing new approaches to supply chain management of oil, gas and mines that consider the challenges associated with new global economic threats like COVID-19—and other pandemics that are yet to be raised. Because of the oil price shock of 2020, government intervention could lead to the reconsideration of developing or improving a local supply chain—because insuring against global health risks. In other cases, the results may advise a sourcing strategy that aims to open up a more flexible supply of goods and services from multiple locations; it may be that multiple suppliers are now less expensive to manage and adapted to new global health risks than to streamline procurement through one major supplier

On the other hand, Morris states that "international companies are depicted as large, integrated firms characterised by strong, centralised coordination from headquarters in western capitals".[41] This means that

[40] Hansen, M.W., 2014. From Enclave to Linkage Economies? A Review of the Literature on Linkages Between Extractive Multinational Corporations and Local Industry in Africa (No. 2014: 02). DIIS Working Paper.

[41] Morris, M., Kaplinsky, R. and Kaplan, D., 2011. Commodities and Linkages: Industrialisation in Sub Saharan Africa. MMCP discussion paper; no. 13.

international companies operate in an enclave manner. Hence it is difficult for the local companies to integrate the oil and gas value chain. This triggered the need for governments to be stricter in dealing with foreign firms. This is what brings about the adoption of local content policies in an oil and gas industry. The integration can be both using public and private inputs. By government developing targeted programmes to enhance local supplier capacity, acquire market information. Governments can have a joint venture with an international company.[42] To increase linkage potential, gain spill-overs and facilitate the technological transfer. The choice of the government to intervene in the oil and gas industry or have a passive attitude towards the industry is entailed in-state participation. Different governments embrace different forms of state participation due to their unique objectives and circumstances.[43] These are known as full equity participation, carried equity participation and "free" equity participation.

Full Equity Participation

In full equity participation, the government invests through its own state-owned companies or small-medium enterprises, such as the national oil companies. The government can invest together with the private entity either by acquiring a majority or minority interest in a joint venture.

Carried Equity Participation

This may take numerous forms. There can be a partial carry of the state's equity in a joint venture. The state may opt to be liable in the first stages of the oil and gas operations. Later the state's equity can be transformed into full equity participation. "A full carried interest is when all

---

[42] Hoekman, B. and Martin, W., 2012. Reducing Distortions in International Commodity Markets: An Agenda for Multilateral Cooperation.

[43] McPherson, C., 2010. 9 State Participation in the Natural Resource Sectors. The Taxation of Petroleum and Minerals: Principles, Problems and Practice, pp. 263–288.

the incurred costs are liable by the private entity".[44] The compensation is paid out of the project invested.

"Free" Equity Participation

This is the equity given to the state from the private entity without any obligations thereof. The state receives the equity interest in the oil and gas operators as a grant from the private entity.

2.5 Government Intervention

Government intervention in the oil and gas sector is necessary due to the distinctive economic nature and institutional requirements of the industry.[45] Hausmann and Rodrick state that "governmental intervention in relevant industries entails choosing favourites or some would say picking winners".[46] In Norway during the petroleum authorisation process. The government awarded licenses to foreign companies that had established clear intentions and specified mechanisms for technology transfer and development of the local supply industry.[47] Government intervention is meant to solve industrial bottlenecks. Though because it is "politically driven", the intervention is vulnerable to some irregularities in some countries. This is evident in Angola how local content policies have been used to gain political support.[48]

Government intervention is implementable in one of two ways. Either through market-based inputs or public inputs. Hausmann and Rodrick further state that "there are functions that markets cannot perform, such

---

[44]McPherson, C., 2010. 9 State Participation in the Natural Resource Sectors. The Taxation of Petroleum and Minerals: Principles, Problems and Practice, pp. 263–288.

[45]Hausmann, R. and Rodrik, D., 2006. *Doomed to Choose: Industrial Policy as Predicament*. John F. Kennedy School of Government, Harvard University, 9.

[46]Ibid.

[47]Nordås, H.K., Vatne, E. and Heum, P., 2003. The Upstream Petroleum Industry and Local Industrial Development: A Comparative Study.

[48]Ovadia, J.S., 2014. Local Content and Natural Resource Governance: The Cases of Angola and Nigeria. *The Extractive Industries and Society*, 1(2), pp. 137–146.

as establishing company registries, setting corporate norms, reasonable access to procurement opportunities for local suppliers, import quotas, sector-specific benefits, enforcing contracts and laws, and providing infrastructure".[49] But the failure of the state to deliver these public inputs affects the efficiency of local content policies. The type of intervention targeting the oil and gas sector must be aligned to other policies such as the industrial policy, the procurement policy and the depletion policy. This must be considered in line with the unique circumstances of a given country. Government intervention occurs throughout the energy lifecycle to facilitate local integration. Jenkins, McCauley and Forman state that such decisions of the government to intervene the energy lifecycle as well as "the design and construction of new facilities, the transport of materials, and their safety and security, as well as issues of common international concern"[50] must all explicitly represent the principles of distributional justice, justice as recognition and procedural justice.

The importance of energy justice in government intervention is to use it as a tool that can facilitate cracking through the enclave of the extractive sector. To be "just and fair", thus contributing to minimising injustices that are brought about by the extractive industry.[51] Government intervention aims at maximising local content efficiency within its energy industry. This entails developing a well-defined local content policy approach while reaching a level of equitable supply of opportunities to the local population at a favourable cost. Energy justice, therefore, determines what should entail a well-defined local content policy. The development of the policy and the cost at which this development is being attained. Since the extractive industry is a capital-intensive industry that requires sophisticated goods, services and technology, such necessities of any industry make it hard for the impoverished local populations to integrate the extractive industry lifecycle thus creating an

[49] Hausmann, R. and Rodrik, D., 2006. *Doomed to Choose: Industrial Policy as Predicament*. John F. Kennedy School of Government, Harvard University, 9.

[50] Jenkins, K., McCauley, D. and Forman, A., 2017. Energy Justice: A Policy Approach. *Energy Policy*, 105, pp. 631–634.

[51] Fuller, S. and McCauley, D., 2016. Framing Energy Justice: Perspectives from Activism and Advocacy. *Energy Research & Social Science*, 11, pp. 1–8.

enclave for the locals. Such hard ordained fact create negative sentiments from the local towards foreign investors. Energy justice has been used as a tool to guide the design of the local content policy to break through the enclave of the extractive sector. Hence energy justice repairs the harm committed by the energy industry as well as repairing the harm done. Energy justice has transformed its principles into policy and industry practice through Environmental Impact Assessments (EIA), Social License to Operate (SLO) as well as the Energy Finance Reserve Obligation (EFRO).[52]

2.6 Benefits of Local Content Policies

In reviewing the literature, there are five significant benefits of local content policies.

2.7 Introduction

Local content is a double-edged sword characterised by both advantages and disadvantages. The policy's efficiency and functionality are dependent on the nature of the oil and gas industry as well as the governing structure of the industry within a particular country. In countries with weak governance associated with destructive patronage politics, corruption, low economic growth and an unstable business climate, local content policies cannot thrive.[53] The following are the advantages and disadvantages of local content policies.

I. Facilitate Cluster Formation and Regional Trade Synergies:

Clusters can be created and facilitated by local content policies. Clusters encourage the transfer of technology which results in high rates

[52]Heffron, R.J. and McCauley, D., 2017. The Concept of Energy Justice Across the Disciplines. *Energy Policy*, 105, pp. 658–667.

[53]Boldbaatar, D., Kunz, N.C. and Werker, E., 2019. Improved Resource Governance Through Transparency: Evidence from Mongolia. The Extractive Industries and Society.

of innovation by creating competent channels to public goods and other market determinants that would have been expensive to acquire. Norway established local supplier clusters that were involved in the oil and gas industry.[54] These established clusters in Norway employed approximately 114,000 people in the local market and sales worth $32 billion in the local economy.[55] Local content policies through cluster creation can reduce the expenditure margins on R&D and improve productivity. Some countries have brought about a regional cluster development. Whereby clusters function beyond the domestic market. This is by developing regional trade synergies through regional integration within a given geographical region.[56] In its 2010 Economic Transformation Program, Malaysia established mechanisms to enable Malaysia into becoming a regional hub for the oil field services and equipment (OFSE) industry.[57]

II. Diversify The Economy:

In many countries with natural resource wealth, there is low economic development. This is due to what is known as a resource curse. Economies with market failure can be negatively affected by the volatility of commodity prices. Humphreys states that "this could lead to an ineffective specialisation in non-tradable goods".[58] Local content policies facilitate the diversification of the economy. This is possible through public based and market-based inputs. That create development channels for other industries other than the oil and gas industry. For example, in Trinidad and Tobago the government through their local content policy. Chose to create incentives for inward investments and joint

---

[54]Sasson, A. and Blomgren, A., 2011. Knowledge Based Oil and Gas Industry. Research Report 3/2011. BI Norwegian Business School. Department of Strategy and Logistics.

[55]Ibid.

[56]Tordo, S. and Anouti, Y., 2013. *Local Content Policies in the Oil and Gas Sector: Case Studies.* World Bank Publications.

[57]Economic Transformation Program, 2010 ("ETP") The ETP Embodies the Government Project to Transform Malaysia into a High-Income Country by 2020. Malaysia Aims to Become a Regional Hub by 2017 and Private Sector Plays a Crucial Role.

[58]Humphreys, M., et al., 2007. *Introduction in "Escaping the Resource Curse".* Columbia University Press.

ventures between local and foreign market participants.[59] This facilitated the capital-linkage from the oil and gas industry to other industries with limited domestic competitiveness. Equally, through their 2010 Economic Transformation Plan, Malaysia adopted local content policies in the same design, facilitating economic diversification.

III. Linkage Promotion in the Domestic Economy:

Linkages may take any of the channels such as upstream linkages, downstream linkages, horizontal linkages, consumption linkages, knowledge and technology linkages and spatial linkages.[60] A country can adopt appropriate local content policies to foster linkages from the foreign direct investment to the domestic economy. These policies facilitate the financial inclusion of small-medium enterprises (SMEs). This helps SMEs gain from preferential supplier programmes. Local content policies can access opportunities beyond domestic borders. Uganda's Local Content Bill of 2017 advocates for the procurement of goods and services in neighbouring countries.[61] If the goods and services being procured cannot be attained within Uganda's domestic economy. This will facilitate linkage spill-over between Uganda and the neighbouring countries.

[59] IPIECA (International Petroleum Industry Environmental and Conservation Association), 2011. "Local Content Strategy: A Guidance Document for the Oil and Gas Industry." Prepared by the International Petroleum Industry Environmental and Conservation Association.

[60] Linkages to the Resource Sector. The Role of Companies, Government and International Development Corporation. Upstream linkages are known as linkages that are related to the procurement of goods and services that the oil and gas industry needs. The downstream linkages, these are result from the further processing of the extracted commodity (also known as forward linkages). Downstream linkages are considered to be expensive and they are not emphasised in Sub-Saharan African local content policies. Consumption linkages are a result of the increased spending due to the earnings attained from the oil and gas industry. The knowledge and technological linkages, result from the transfer of knowledge and technical know-how from the foreign investor to the local companies. Spatial linkages are benefits from shared infrastructural projects.

[61] Uganda The Local Content Bill, 2017.

IV. Addressing Market Failure:

The discovery of natural resources increases government revenues. The industry develops into a large percentage of the domestic economy's GDP.[62] Paul Collier states that "if the economy does not have the absorptive capacity to make efficient use of these revenues, the result can be inflation or exchange rate appreciation".[63] This brings about Dutch disease. Local content policies have been used to create the absorptive capacity of a domestic economy. In Norway, the government emphasised the development of the national oil company Statoil.[64] The Norwegian government was of the assumption that Statoil would facilitate the development and inclusion of local suppliers as well as avail control of the local oil and gas industry. The government also created capacity through the country's expertise in shipbuilding and marine services. Foreign investors are predominantly big corporations that displace local companies from the local oil and gas market. Local content policies adopt minimum targets that "must" be met for the inclusion of local companies. This preferential inclusion brings about supplier development in the domestic industry.

V. Provision for Social Objectives:

The oil and gas industry calls for a sophisticated and highly skilled labour force. Resource-rich countries like Nigeria, Angola, Tanzania and Uganda do not have the required skills. Local content policies create mechanisms for local employment. In Brazil, local content policies involved the national oil company Petrobras in skills development. Petrobras created a human resource programme through the Agencia Nacional

---

[62] Humphreys, M., et al., 2007. *Introduction in "Escaping the Resource Curse"*. Columbia University Press.

[63] Collier, Paul, and Venebles, Anthony, 2011. Plundered Nations? Successes and Failures in Natural Resource Extraction. Palgrave Macmillan.

[64] Dutch disease is the negative impact on an economy of anything that gives rise to a sharp inflow of foreign currency, such as the discovery of large oil reserves. The currency inflows lead to currency appreciation, making the country's other products less price competitive on the export market.

do Petroleo in 1999.[65] The programme encouraged training courses that met the demands of the oil and gas industry. Local content policies give international oil companies a "license to operate".

2.8 Disadvantages of Local Content Policies

I. State Fragility:

Paul Collier states that "there are three main reasons why resource abundance can threaten the quality of governance and create an increased risk of state fragility. These are looting due to the increased incentives for corruption. Since the government has proceeded from the oil and gas industry. Reduction of accountability, since resource revenue may reduce citizen scrutiny of government. Presence of a difficult economic environment is due to the extreme volatility of the political climate".[66] Local content policies are a contributing factor to state fragility. Because powerful factions use these policies to gain political support. This is known as neopatrimonialism.

The two Presidents with the longest tenure in Africa are; President Obiang of Equatorial Guinea and President Dos Santos of Angola. These two Presidents have ruled their respective countries since 1979. By 2015 Angola was producing 1.8 million (BPD) making the country the biggest producer of crude oil in Sub-Saharan Africa.[67] Equatorial Guinea follows from Angola as the third biggest producer of crude oil in the region, and the oil-rich country has more than 1.1 billion barrels in crude oil.[68] McSherry states that "the neo-patrimonial nature of Equatorial Guinea's regime, however, makes it likely that any attempts by Obiang to quell opposition will take the form of greater spending on clientelism,

[65] Mendonça, R.W. and de Oliveira, L.G., 2013. Local Content Policy in the Brazilian Oil and Gas Sectoral System of Innovation. *Latin American Business Review*, 14(3–4), pp. 271–287.

[66] Collier, Paul, and Venebles, Anthony, 2011. *Plundered Nations? Successes and Failures in Natural Resource Extraction.* Palgrave Macmillan.

[67] McSherry, B., 2006. The Political Economy of Oil in Equatorial Guinea. *African Studies Quarterly*, 8(3).

[68] Ibid.

patronage and granting of contracts, rather than widespread benefits to the masses".[69] Such political settlements, practise exclusionary politics in developing the oil and gas as well as hindering inclusive development. Researchers such as Tomas Frederiksen have found that even corporate responsibility practices can work hand in hand with local content policies to entrench clientelist political settlements like the one in Equatorial Guinea.[70]

II. Lack of Clarity:

Local content policies have been noted to lack clarity in their definition. Developing countries that have adopted local content policies they do not ascertain what the policy entails and in which manner the policy will be implemented. When adopting local content policies, most policymakers do not tailor these policies according to the circumstances in the country. This makes the policy ineffective as it creates hugely unrealistic expectations. Lastly, the lack of monitoring and enforcement mechanism is also a significant detriment to the development of local content policies. Countries like Tanzania, Kenya and Uganda have numerous state agencies regulating, defining and setting local content requirements for the same companies in the country. This creates confusion on who has the mandate to enforce, to monitor and to facilitate the development of local content policies.

III. Redundancy:

The minimum targets placed by these policies would be undertaken even without the adoption of such policies. Foreign companies will "almost always" seek local suppliers because it gives them a social license to operate. In most occasions, this is included within a company's corporate social responsibility. Local content policies can deter FDI into an economy. This is due to the strict local content requirements in the

---

[69] Stephenson, Sherry, 2013. Addressing Local Content Requirements: Current Challenges and Future Opportunities. *Biores*, 7(3), 25 July.

[70] Frederiksen, T., 2019. Political Settlements, the Mining Industry and Corporate Social Responsibility in Developing Countries. *The Extractive Industries and Society*, 6(1), pp. 162–170.

domestic economy. Hufbauer, Schott and Cimino state that "the effect of local content policies on trade is to discourage imports and to stifle competition between domestic and foreign firms. The impact on trade of local content policies depends on the percentage of local content required and the efficiency of existing firms. In an economy with inefficient firms, a high degree of required local content thwarts competition".[71] Furthermore, Stephenson states "while proponents argue these policies are a short-term policy, put in place to protect infant industries and businesses. Opponents point out that once local content policies become a mainstay, withdrawal of government support will often be met with fierce resistance. It is also possible that the relevant manufacturing sectors will never attain the level of efficiency necessary to operate without government support, and instead require continuous government support".[72]

IV. Rent-Seeking:

The oil and gas industry creates a large part of a country's GDP in resource-rich countries. These policies bring the expectation for local inclusion. Hence the labour force of resource-rich countries seeks significant proceeds from the industry. Rent-seeking wastes competition for a given economic opportunity and in this case, the opportunity is the oil and gas industry.

2.9 Research Questions

I. Why can't national content policies break through the enclave of the oil and gas industry in developing countries? Can energy justice be used as a tool for creating "just-mechanisms" in adopting local content policies?
II. How can regional integration be instrumental in the creation of local content policies?

[71] Hufbauer, G.C., Schott, J.J. and Cimino-Isaacs, C., 2013. *Local Content Requirements: A Global Problem* (Vol. 102). Columbia University Press.

[72] Stephenson, Sherry, 2013. Addressing Local Content Requirements: Current Challenges and Future Opportunities. *Biores*, 7(3), 25 July.

The main objective of the research questions is to explore, study and avail solutions for the inefficiency of local content policies in developing countries such as Nigeria, Angola, Tanzania, Kenya and Uganda. The first question attempts to understand as to why national content policies cannot meet the demands of the oil and gas industry in developing countries, unlike in developed countries such as Norway, Brazil, and United Kingdom. For instance, in Norway and Brazil, the economies relied heavily on the shipping industry that could be leveraged to develop skills for the offshore oil and gas resources. But in Nigeria and Angola, the main driver of the economy was subsistence farming, the skills in this sector cannot even be adopted to service the oil and gas industry. Hence different attributes will be used to explore as to why national content policies fail. Whether the answer relies solely on politics (such as democracy, the rule of law, stable economic environment, etc.) Or technicalities of the demanding oil and gas industry.

Furthermore, the question brings in energy justice as a tool that can be used to create a just-mechanism in adopting and implementing local content policies in developing countries. Energy justice is used to construct innovative means to crack through the enclave of the extractive sector. Local content policies have fundamentally been adopted without any guiding mechanism. These policies have been implemented based on the desired outcome of integrating the local population within the oil and gas lifecycle. Energy justice, through its tenets, create a mechanism of adopting local content policies while recognising all affected and marginalised parties, while following the right procedures while ensuring the equitable distribution of ills and benefits accordingly? Since the oil and gas industry is a capital-intensive industry that calls for stringent prerequisites that may often not be readily available for the general local population

The last question mentions the role of regional integration in adopting local content policies. Literature shows that developing countries cannot efficiently reap the benefits of their oil and gas resources due to the numerous bottlenecks that tackle the capability to develop individually as a country. Thus countries that are near, within the same regional economic bloc, bearing similar natural resources such as Kenya, Tanzania and Uganda or Mozambique, Angola and Nigeria. Have a unique opportunity as nascent oil and gas-rich countries to create a regional integration approach through the East African Community or the Economic Community of West African States and create a regional content approach for the regional economic blocs.

2.10 Contextual Aspect of Developing Local Content Policies

In Sub-Saharan rich extractive countries such as Uganda, Tanzania, Kenya, Angola and Nigeria, low levels of governance and accountability persist, which adversely perpetuates nepotism and partisan natural resource governance that creates an environment conducive for enclave development. This section of the book aims to highlight further the challenges of designing credible optimum local content policies in Sub-Saharan oil and gas-rich countries. One would ask if the plans undertaken by the oil and gas companies in designing local content policies materialise into meaningful action, or do they favour the few with no fruition for the local and direct community having resources exhausted in their backyard?[73] Little is known about how the management of oil and gas operations comes about designing the local content policies that are mostly transplanted from developed economies such as Norway, Brazil and the United Kingdom. These policies that are often seen as "cutting edge" in local content reports, websites and company brochures, and more specifically how these oil and gas companies arrive at decisions on what programme to undertake, whom the programme targets for what reasons and how these initiatives are undertaken and in what setting.[74] This is of paramount importance because oil and gas companies come to invest with intact linkages from foreign countries with financial and technical muscle to protect the enclave nature of the oil and gas industry. Furthermore, the compatibility of these local content

[73]Alzina, Virginia et al., 2004. Enclave Projects? What Are They and What Can Be Done? Document of the Inter-American Development Bank.

[74]Ackah-Baidoo, A., 2012. Enclave Development and 'Offshore Corporate Social Responsibility': Implications for Oil-Rich Sub-Saharan Africa. *Resources Policy*, 37(2), pp. 152–159.

plans and programmes that are embedded within the company's budget books and modus operandi when it comes to local content. How far are they in line with the objectives and targets by the regulators and policy-makers of a given oil and gas-rich developing country? This is why this book brings in the aspect of energy justice in local content design and development due to the enclave nature of the oil and gas industry. Most literature offers minimal guidance on this front.

Looking back at Ferguson, there is probably no industry in Sub-Saharan Africa that has developed more enclave-like fashion in its operation than the oil and gas sector.[75] Nascent oil and gas regions like the East African region boast the possibility of East African economies gaining massive extractive FDI and profits from their natural resources, unlike their West African counterparts. All this presumptuous promise is dependent on the fact that the resources be managed appropriately. Furthermore, the revenues flowing from this nascent lucrative sector could mitigate some of the economic woes plaguing these oil and gas-rich East and West African countries.[76] Most researchers have held that currently, the revenues obtained from royalties, fiscal frameworks, work programmes and cash-stimulus cheques are not being used positively for development, due to the enclave nature of the industry and the lack of synergy between the private and public sector. Foreign oil and gas companies and the local population find themselves dealing with local content policies that are mostly nationalistic, causing investor-flight as well as building a local elite capture of the industry through nepotism and clientelism. Given the uniqueness of this operating environment, it is clear that local content policies design must be dynamic and adopt an energy justice framework for these policies to be effective. The simple mechanisms of achieving local content targets that have been set by regulators without prior negotiations and consent of the different stake-holders within the oil and gas value chain.[77] Can no longer function as a manner in which local content policies are designed and adopted

[75]Ferguson, James, 2005. Seeing Like an Oil Company: Space, Security, and Global Capital in Neoliberal Africa. *American Anthropologist*, 107(3), pp. 377–382.

[76]Ibid.

[77]Ovadia, J.S., 2017. Local Content in Tanzania's Gas and Minerals Sectors: Who Regulates? CMI Brief.

to gain any significant fruition from the policy and the investment as a whole. Oil and gas companies must go beyond what is promulgated in the local content regulations to attain a justified inclusive oil and gas industry in any region. The view is that it is hypocritical for an oil and gas company to agree to meet local content provisions that it knows will never materialise due to the nature of the host countries such as Kenya, Uganda, Tanzania and Nigeria. Countries that do not have the absorptive capacity and technical know-how in the population to meet the demands of the local content provisions.[78] This breeds seeds of corruption and mal-practise while icing out the local community due to the enclave nature of the oil and gas industry.

The rapid increase in international interest in oil in African regions such as East and West Africa is part of growing excitement about the region's oil and gas resource. Hence a "new frontier" for industry and multinational corporations: it is being encouraged today by a range of business enthusiasts. Recent high-profile trips by US Presidents George Bush and Barack Obama to the African Continent have opened the investors "eyes to the region's economic growth". Former US vice president Dick Cheney noted that West Africa, in particular, could become one of America's fastest-growing oil and gas sources. They are adding that Africa's high-quality oil makes it an attractive alternative refinery to the East Coast of the United States.[79] The fact that enclaves in the oil and gas industry are far more pronounced than those that have taken shape in most other sectors of the extractive industry in Sub-Saharan Africa.[80] Has significant implications for the development of local content, affecting the ability of companies to function in any community and having an inclusive value chain.[81] One of the biggest challenges for any oil and gas producer in this region that embraces local content policies is identifying who the community is. As the recent developments in oil production

[78]Hilson, G. and Maconachie, R., 2009. Good Governance and the Extractive Industries in sub-Saharan Africa. *Mineral Processing and Extractive Metallurgy Review*, 30(1), pp. 52–100.

[79]Pegg, Scott, 2006. Can Policy Intervention Beat the Resource Curse? Evidence from the Chad–Cameroon Pipeline Project. *African Affairs*, 105(418), pp. 1–25.

[80]Hilson, G., 2012. Corporate Social Responsibility in the Extractive Industries: Experiences from Developing Countries. *Resources Policy*, 37(2), pp. 131–137.

[81]Ibid.

in Ghana have clearly illustrated, the majority of drilling activities take place offshore, and tankers load oil at seaports.[82] Who, then, should be channelling revenues to? In theory, unlike other extractive activities such as gold mining (or perhaps even in onshore oil-producing environments such as those in the Niger Delta in Nigeria), a community blueprint needs to be developed and followed because activity takes typically place near human settlements.[83] Such enclave developments are the main reason this book applies energy justice in local content development, as explained in later chapters.

2.11 Summary

This chapter provides a general overview of what local content policies are and what they entail in when adopted and implemented in any oil and gas country. The chapter further illustrates the historical perspective of local content policies. It is only through understanding the origins of local content policies may policymakers and regulators know the value of such policies through an in-depth comparative analysis as one done in this book. The chapter finds that local content policies have been developed from other policies such as import substitution policies of which they are all known as production development policies. As much as local content policies are favourable within an economy, the chapter finds that the policy is also susceptible to corruption, especially in weak democracies and countries with low accountability mechanisms due to poor institutions. They are also policies that favour local integration into an industry as well as the protection of local firms and entities. Such susceptibility of local content policies is what creates an enclave oil and gas industry primarily in countries like Angola, Tanzania, Kenya, Uganda and Mozambique. Having an enclave industry prohibits the development of linkages and growth of other key strategic sectors of the economy. The literature on local content policies always mentions the effects of

[82] Pegram, J., Falcone, G. and Kolios, A., 2020. Job Role Localisation in the Oil and Gas Industry: A Case Study of Ghana. *The Extractive Industries and Society*, 7(2), pp. 328–336.

[83] Ackah-Baidoo, A., 2012. Enclave Development and 'Offshore Corporate Social Responsibility': Implications for Oil-Rich Sub-Saharan Africa. *Resources Policy*, 37(2), pp. 152–159.

governance of the oil and gas industry on the policy. But the chapter finds that local content policy is closely linked to depletion policies. In that, the adequate and appropriate synchronisation of local content policies and the tenets of energy justice such as procedural, recognition and distributive justice would yield favourable results regardless of the nature of the oil and gas industry in both developed and developing oil and gas countries.

References

Ackah-Baidoo, A., 2012. Enclave Development and 'Offshore Corporate Social Responsibility': Implications for Oil-Rich Sub-Saharan Africa. *Resources Policy*, 37(2), pp. 152–159.

Al-Kasim, F., 2006. *Managing Petroleum Resources: The "Norwegian Model" from a Broad Perspective* (Vol. 30). Oxford Institute for Energy Studies.

Alzina, Virginia et al., 2004. Enclave Projects? What Are They and What Can Be Done? Document of the Inter-American Development Bank.

Boldbaatar, D., Kunz, N.C. and Werker, E., 2019. Improved Resource Governance Through Transparency: Evidence from Mongolia. The Extractive Industries and Society.

Campos, J.E. and Pradhan, S. (Eds.), 2007. *The Many Faces of Corruption: Tracking Vulnerabilities at the Sector Level*. World Bank Publications.

Collier, Paul, and Venebles, Anthony, 2011. *Plundered Nations? Successes and Failures in Natural Resource Extraction*. Palgrave Macmillan.

Ferguson, James, 2005. Seeing Like an Oil Company: Space, Security, and Global Capital in Neoliberal Africa. *American Anthropologist* 107(3), pp. 377–382.

Frederiksen, T., 2019. Political Settlements, the Mining Industry and Corporate Social Responsibility in Developing Countries. *The Extractive Industries and Society*, 6(1), pp. 162–170.

Fuller, S. and McCauley, D., 2016. Framing Energy Justice: Perspectives from Activism and Advocacy. *Energy Research & Social Science*, 11, pp. 1–8.

Gidado, M.M., 1999. Petroleum Development Contracts with Multinational Oil Firms: The Nigerian Experience. Ed-Linform Services.

Hansen, M.W., 2014. From Enclave to Linkage Economies? A Review of the Literature on Linkages Between Extractive Multinational Corporations and Local Industry in Africa (No. 2014: 02). DIIS Working Paper.
Hausmann, B.M., Moreira, M. and Rodrik, D., 2006. Coordination Failures, Clusters, and Microeconomic Interventions. Andrés Velasco.
Hausmann, R. and Rodrik, D., 2006. *Doomed to Choose: Industrial Policy as a Predicament*. John F. Kennedy School of Government, Harvard University.
Heffron, R.J. and McCauley, D., 2017. The Concept of Energy Justice Across the Disciplines. *Energy Policy*, 105, pp. 658–667.
Heum, P., 2008. Local Content Development: Experiences from Oil and Gas Activities in Norway.
Hilson, G., 2012. Corporate Social Responsibility in the Extractive Industries: Experiences from Developing Countries. *Resources Policy*, 37(2), pp. 131–137.
Hilson, G. and Maconachie, R., 2009. Good Governance and the Extractive Industries in Sub-Saharan Africa. *Mineral Processing and Extractive Metallurgy Review*, 30(1), 52–100.
Hoekman, B. and Martin, W., 2012. Reducing Distortions in International Commodity Markets: An Agenda for Multilateral Cooperation.
Hufbauer, G.C., Schott, J.J. and Cimino-Isaacs, C., 2013. *Local Content Requirements: A Global Problem* (Vol. 102). Columbia University Press.
Humphreys, M., 2007. *An Introduction to "Escaping the Resource Curse"*. Columbia University Press.
IPIECA (International Petroleum Industry Environmental and Conservation Association).
Jenkins, K., McCauley, D. and Forman, A., 2017. Energy Justice: A Policy Approach. *Energy Policy*, 105, pp. 631–634.
Lall, S., 2013. Reinventing Industrial Strategy: The Role of Government Policy in Building Industrial Competitiveness. *Annals of Economics & Finance*, 14(2).
McPherson, C., 2010. 9 State Participation in the Natural Resource Sectors. The Taxation of Petroleum and Minerals: Principles, Problems and Practice, pp. 263–288.
McSherry, B., 2006. The Political Economy of Oil in Equatorial Guinea. *African Studies Quarterly*, 8(3).
Mendonça, R.W. and de Oliveira, L.G., 2013. Local Content Policy in the Brazilian Oil and Gas Sectoral System of Innovation. *Latin American Business Review*, 14(3–4), pp. 271–287.

Monge-González, R., Rivera, L. and Rosales-Tijerino, J., 2010. Productive Development Policies in Costa Rica: Market Failures, Government Failures, and Policy Outcomes.

Morris, M., Kaplinsky, R. and Kaplan, D., 2011. Commodities and Linkages: Industrialisation in Sub Saharan Africa. MMCP discussion paper; no. 13.

Ngoasong, M.Z., 2014. How International Oil and Gas Companies Respond to Local Content Policies in Petroleum-Producing Developing Countries: A Narrative Enquiry. *Energy Policy*, 73, pp. 471–479.

Nordås, H.K., Vatne, E. and Heum, P., 2003. The Upstream Petroleum Industry and Local Industrial Development: A Comparative Study.

Nwapi, C., 2015. Defining the "Local" in Local Content Requirements in the Oil and Gas and Mining Sectors in Developing Countries. *Law and Development Review*, 8(1), pp. 187–216.

Ovadia, J.S., 2012. The Dual Nature of Local Content in Angola's Oil and Gas Industry: Development vs Elite Accumulation. *Journal of Contemporary African Studies*, 30(3), pp. 395–417.

Ovadia, J.S., 2014. Local Content and Natural Resource Governance: The Cases of Angola and Nigeria. *The Extractive Industries and Society*, 1(2), pp. 137–146.

Ovadia, J.S., 2017. Local Content in Tanzania's Gas and Minerals Sectors: Who Regulates? CMI Brief.

Pegg, Scott., 2006. Can Policy Intervention Beat the Resource Curse? Evidence from the Chad–Cameroon Pipeline Project. *African Affairs*, 105(418), 1–25.

Pegram, J., Falcone, G. and Kolios, A., 2020. Job Role Localisation in the Oil and Gas Industry: A Case Study of Ghana. *The Extractive Industries and Society*, 7(2), pp. 328–336.

Ramdoo, I., 2013. Fixing Broken Links.

Ramdoo, I., 2015. Unpacking Local Content Requirements in the Extractive Sector: What Implications for the Global Trade and Investment Frameworks. By International Centre for Trade and Sustainable Development (ICTSD).

Rodriguez-Clare, A., 2007. Clusters and Comparative Advantage: Implications for Industrial Policy. *Journal of Development Economics*, 82(1), pp. 43–57.

Rodrik, D., 2006. What's So Special About China's Exports? *China & World Economy*, 14(5), pp. 1–19.

Sasson, A. and Blomgren, A., 2011. Knowledge-Based Oil and Gas Industry. Research Report 3/2011.

Stephenson, Sherry, 2013. Addressing Local Content Requirements: Current Challenges and Future Opportunities. *Biores*, 7(3), 25 July.

Stiglitz, J.E. and Dasgupta, P., 1981. Resource Depletion Under Technological Uncertainty.
Tordo, S. and Anouti, Y., 2013. *Local Content Policies in the Oil and Gas Sector: Case Studies*. World Bank Publications.
Tordo, S., 2011. *National Oil Companies and Value Creation*. The World Bank.
Uganda The Local Content Bill, 2017.
Victor, N.M., 2007. On Measuring the Performance of National Oil Companies (NOCs). Program on Energy and Sustainable Development working papers, WP64, Stanford University.

3

Local Content and the Energy Justice Framework

3.1 Introduction

Oil and gas have been vital in the development of economies in both developing and developed regions. The energy sector has played a pivotal role in employment generation, economic development, creation of industries and creation of wealth for nations.[1] The presence of oil and gas brings about the adoption and implementation of energy law. Heffron states that "the importance of energy law and environmental law is in providing legislation to manage the natural resources of a country, and in their potential for changing human and societal behaviour; another important characteristic is that of policy formulation".[2] Though energy law manages the changing human and societal behaviour towards oil and gas. But human behaviour alone is not the primary driver of energy law. Energy law has five main drivers, namely: safety, security, economics, infrastructure and justice.[3] These drivers are fundamental in creating the

[1] Heffron, R.J., 2014. *Energy Law: An Introduction.* Springer.

[2] Ibid.

[3] Heffron, R.J. and Talus, K., 2016. The evolution of energy law and energy jurisprudence: Insights for energy analysts and researchers. *Energy Research & Social Science*, 19, pp. 1–10.

R. S. Muhongo, *Energy Justice*, Energy, Climate and the Environment,
https://doi.org/10.1007/978-3-030-61338-9_3

foundation for local content policies to crack through the enclave nature of the oil and gas industry.

Oil and gas have an energy lifecycle, and the energy lifecycle starts from extraction, production, operation and supply, consumption and finishes at the waste management stage.[4] The energy lifecycle does not exist mechanically; instead, there is a human element that engages in the creation of every stage of the energy lifecycle. This human element can also be determined as societal preferences. The changes in societal preferences encourage the different drivers of energy law in different jurisdictions dependant on the policy maker's intentions. We are putting together societal preferences, oil and gas and the energy lifecycle. It is essential to define energy law "as the regulation of energy-related rights and duties of various stakeholders over oil and gas over the energy lifecycle".[5] Energy law is also essential for policy formulation, as much as energy law and policy is concerned with markets, the security of supply and efficiency. Justice as a driver of energy law is what will be used to create a theoretical framework that offers policy guidelines on how to crack through the enclave of the oil and gas industry, especially in developing countries. The local content policy is a production development policy, and this would call for the theoretical framework of such a policy be built under resource law and policy rather than energy law and policy.[6] But justice is a driver of energy law and not resource law. The two-tier principles of energy justice, the first being the triumvirate of tenets—distribution, procedural and recognition of justice.[7] The second tier will be the eight-core principles, namely: availability, affordability, due process, transparency and accountability, sustainability, intra-generational equity, inter-generational equity and responsibility.[8] A

[4]Ibid.

[5]Heffron, R.J. and Talus, K., 2016. The evolution of energy law and energy jurisprudence: Insights for energy analysts and researchers. *Energy Research & Social Science*, 19, pp .1–10.

[6]This is because resource law and policy deals with development policies, industrial policies as well as maximisation of government revenue.

[7]Heffron, R.J. and McCauley, D., 2017. The Concept of Energy Justice Across the Disciplines. *Energy Policy*, 105, pp. 658–667.

[8]Sovacool, B.K., Heffron, R.J., McCauley, D. and Goldthau, A., 2016. Energy Decisions Reframed as Justice and Ethical Concerns. *Nature Energy*, 1(5), p. 16024.

key reason as to why justice is needed in the adoption and implementation of local content policies is the susceptibility of having corrupt local content policies in an enclave oil and gas industry. That leads to catastrophic results such as having a resource curse in countries like Nigeria, The Democratic Republic of Congo and Angola. Hence to ensure fairness and equality in society, in essence, resolve inequalities, the book employs a justice framework throughout the local content adoption and implementation.

3.2 Energy Justice and Local Content Policies

Energy justice is a new interdisciplinary phenomenon that aims to apply principles of justice in energy policy, energy production and systems, energy consumption, energy activism, energy security and climate change.[9] The principles of energy justice will be extended to the adoption of local content policies in this book, as a means of identifying the present injustices, extending the definitions what "local" means in local content by developing new approaches towards local content policy implementation and lastly availing remedy options for jurisdictions that have already adopted unfavourable local content policies in their energy lifecycle. Justice is a means of ensuring and recognising that all human beings are equally worthy to the distribution of good and bad things.[10] As much as energy justice is to be used in understanding the distribution of ills in adopting and implementing local content policies. It is important to note that justice in energy emerged due to environmental pollution that brought about environmental justice.[11]

The foundation of energy justice is on the principles of environmental justice. The notion of environmental justice was developed with

---

[9]Jenkins, K., McCauley, D., Heffron, R., Stephan, H. and Rehner, R., 2016. Energy Justice: A Conceptual Review. *Energy Research & Social Science*, 11, pp. 174–182.

[10]McCauley, D.A., Heffron, R.J., Stephan, H. and Jenkins, K., 2013. Advancing Energy Justice: The Triumvirate of Tenets. *International Energy Law Review*, 32(3), pp. 107–110.

[11]Ibid.

regard to the unequal distribution of environmental ills (such as environmental pollution).[12] Local content policies are developed for the sole purpose of creating channels for local participation as well as value addition of the raw oil and gas being extracted. Energy justice aims to avail all the individuals across all areas, with safe, affordable and sustainable energy[13]—the principles of energy justice advocate for the synergy of energy policy across the whole energy system. The energy policy is adopted and implemented to address "the unequal distribution of ills on the whole energy lifecycle".[14] The merge between energy law and local content is because energy policy also advocates for social justice. Local content is a means of attaining social justice to the local population due to the presence of oil and gas. Social justice goes beyond the current government and industry concerns with regard to energy security, economic development and technology as a means of taking into account morality while making decisions.[15]

The morality consideration in decision making, with regard to oil and gas, does not create a fair distribution of ills in the energy industry—local content policy though applicable in the regional, national and local spheres of energy resource governance. The injustices suffered by the affected or marginalised parts of society due to local content policy will not be remedied by social justice.

The theory of energy justice will ascertain the distribution of dividends and ills of the policy; the theory will further show who the local content policy benefits in the energy lifecycle whether the policy benefits the local population close to the oil and gas through the community content approach. Or whether the policy aims to benefit the whole nation that has the oil and gas through the national content policy and whether the policy follows a more cosmopolitan approach of engaging the whole region that has oil and gas through a regional content

---

12 McCauley, D.A., Heffron, R.J., Stephan, H. and Jenkins, K., 2013. Advancing Energy Justice: The Triumvirate of Tenets. *International Energy Law Review*, 32(3), pp. 107–110.

13 Jenkins, K., McCauley, D. and Forman, A., 2017. Energy Justice: A Policy Approach. *Energy Policy*, 105, pp. 631–634.

14 McCauley, D.A., Heffron, R.J., Stephan, H. and Jenkins, K., 2013. Advancing Energy Justice: The Triumvirate of Tenets. *International Energy Law Review*, 32(3), pp. 107–110.

15 Heffron, R.J., Rønne, A., Tomain, J.P., Bradbrook, A. and Talus, K., 2018. A Treatise for Energy Law. *The Journal of World Energy Law & Business*, 11(1), pp. 34–48.

approach. The three typologies of local content need to be assessed if they have been adopted through a correct and righteous procedure that recognises all injustices and sectors of the society. This assessment can only be done through energy justice three core themes distributional justice, procedural justice and recognition justice.[16]

3.3 Socio-Legal Snapshot of Local Content Policies

The local content policy has been implemented as a measure to boost economic development in numerous resource-rich countries from the developed world to the developing world. These policies have been instrumental in driving economies to prosperity. But the implementation and adoption process has been a transplanted mechanism from developed countries to developing countries. This might be attributed to colossal donor dependence. But the results have not been similar in this regard. The two mechanisms of legal transplantation, namely the "copy-paste" method that purely transplants and copy-pastes the legal doctrines and framework of one jurisdiction to another. Secondly, the synchronisation-harmonisation process that goes beyond the copy-paste act. The second mechanism is an actual transition process that adapts the whole legal framework into the local legal culture then merges the legal principles into the local institutional design progressively as more models are imported.

Developing resource-rich countries are fond of adopting legal doctrines from developed countries. This is mostly due to donor dependence, international consultants and intellectual dependence on developed countries. Tanzania, Kenya, Angola, Nigeria and Uganda have adopted a national content policy similar to Norway, and Brazil. Such transplantation of the national content policy has been seen when Nigeria adopted similar policies after the discovery of their oil and gas resources with the help of NORAD (a Norwegian NGO). The case study

[16]Heffron, R.J., Rønne, A., Tomain, J.P., Bradbrook, A. and Talus, K., 2018. A Treatise for Energy Law. *The Journal of World Energy Law & Business*, 11(1), pp. 34–48.

examines as to whether other nascent developing oil and gas producers are prone to similar results as Nigeria due to the transplantation of the national content policy. This examination will assess the political, institutional, social and cultural design of these countries. At the same time, they are assessing the state of these economies before the oil and gas resources. The major hurdle is whether Kenya, Uganda and Tanzania will yield better results from the national content policy.

3.3.1 Distributional Justice

The first tenet of energy justice is distributional justice; Walker thinks that energy justice includes the unequal allocation of environmental benefits and ills as well as the uneven distribution of the responsibilities that are associated thereof.[17] Distributional justice also advocates for the distribution of benefits and ills within the energy system. Local content policies are applicable holistically in the energy systems because from exploration to waste management in energy lifecycle, and local integration is needed to break the enclave nature of the oil and gas industry.

When adopting and implementing local content policies, a decision tree is usually adopted to this mandate. The first step in the decision tree is involved with identifying the country's objectives. In this stage of the local content decision tree, the policy should be adopted along with the different objectives of different stakeholders, and the expectations of these stakeholders should be taken into account.[18] With regard to the ideal implementation of local content policies in resource-rich countries call for mapping out the different stakeholders involved and their objectives towards the oil and gas. But principles of distributional justice can allocate the uneven distribution of ills and benefits in resource-rich countries when it comes to adopting and implementing local content policies

[17]Walker, G., 2012. *Environmental justice: Concepts, Evidence and Politics*. Routledge.

[18]Valerie Marcel and others, 2016. A Local Content Decision Tree for Emerging Producers (Energy, Environment and Resources Department Chatham House: The Royal Institute of International Affairs).

in the energy lifecycle. Oil and gas are usually found in remote, undeveloped regions of a resource-rich country. Local content policies have practically favoured the political elite instead fairly distributing the benefits of the oil and gas. Local communities in resource-rich countries have suffered the ills of the presence of oil and gas rather than their benefits. Local content policies have been used by the political elite to use patronage as a means to secure political loyalty.

3.3.2 Procedural Justice

Procedural justice advocates that all stakeholder within an energy system should be awarded and follow equitable procedures. Hence all stakeholders should be engaged in the energy systems in a non-discriminatory manner.[19] The second step in the local content decision tree regards to the access to information on the resource potential and the evaluation of how the economy can absorb foreign direct investment. This step of the local content decision tree, there is a need to avail information on the national skill base, information on the industries that need support as well as the relevant infrastructure that is needed for the adequate development of oil and gas. Procedural justice advocates that in such a process, the public must be involved in decision making after been given full information on the oil and gas. Davies states that "procedure justice requires participation, impartiality and full information disclosure by the government and industry accompanies by appropriate and sympathetic mechanisms".[20]

In adopting and implementing local content policies, most resource-rich countries (especially in developing nations) have not been able to avail and disclose information to the public readily. Furthermore, the lack of reliable information on the available competencies within a country or region makes the multinational corporations seek outside sources since they do not know the capabilities and competence of the local suppliers.

[19]Heffron, R.J., Rønne, A., Tomain, J.P., Bradbrook, A. and Talus, K., 2018. A Treatise for Energy Law. *The Journal of World Energy Law & Business*, 11(1), pp. 34–48.

[20]Todd, H. and Zografos, C., 2005. Justice for the environment: developing a set of indicators of environmental justice for Scotland. *Environmental Values*, 14(4), pp. 483–501.

This makes local content policies fail to take into account the long-term objectives of the economy. Due to the lack of information, it is difficult for the energy industry within a country to bring about industrialisation, economic diversification, and even strengthen value creation.

3.3.3 Recognition Justice

Recognition justice entails that people must be availed all political rights, as well as being free from physical threats.[21] Recognition justice is undermined in many forms through cultural and political domination, insults, degradation and devaluation.[22] The recognition injustice generally entails that there must be a recognition of societies norms and principles regardless of their social, cultural, ethnic, racial or gender differences.[23] Recognition justice in the local content decision tree should be found in the strategic plan to build capacity to meet the demands of the energy sector as well as who does the monitoring and review of the local content policy within an energy resource-rich country. Recognition justice in a local content decision tree enhances the recognition of marginalised communities and groups in the oil and gas lifecycle.

When developing local content policies, the recognition of different capacity gaps within different communities-facilitates the increase in coordination and efficiency of the local content policy. Most developing countries have adopted the national content approach towards the local content policy. This approach usually neglects the communities that have the oil and gas in their vicinity and concentrate on the metropolitan areas rather than the remote host communities. The community content approach towards local content also emphasises the community is having the oil and gas rather than the rest of the host country. National and community content approach do not meet the standards of recognition

[21] Schlosberg, D., 2003. The Justice of Environmental Justice: Reconciling Equity, Recognition, and Participation in a Political Movement. *Moral and Political Reasoning in Environmental Practice* (Vol. 77, p.106).

[22] Fraser, N., 2009. Social Justice in the Age of Identity Politics. *Geographic Thought: A Praxis Perspective* (pp. 72–91).

[23] Ibid.

justice as other parts of the community are neglected due to the lack of having oil and gas or the unavailability of skills, capital and competence.

3.4 The Implementation of Energy Justice in Local Content Policy Adaptation and Implementation

The triumvirate of tenets of energy justice, namely distribution, procedural and recognition justice is applicable in local content policy through the adaptation of restorative justice. It should be noted that restorative justice is the only means that energy justice input into practice.

The local content policy decision tree starts with assessing the national vision of the economy. To analysing and assessing the energy resource potential, to evaluating the absorptive capacity of the economy, filling the gap to meet the demands of the energy sector that can be met by the local capacity development lastly the monitoring and review as the enforcement segment of the decision tree. Developing countries due to political capture, the lack of capital and technology, corruption and the inability of the local suppliers to meet the demands of the sophisticated energy sector. The local content policy has been characterised by massive irregularities such as rent-seeking, and local content policies have been used for political patronage. Local content policies are costly for developing countries to implement this enhances fronting, where locals use corrupt means to represent foreign multinationals. Such tendencies have created injustices for the host communities, especially in developing countries.

The injustices brought by local content policies need to be attended, and a resolution needs to be met for the adequate adoption and implementation of these policies. The only manner in which energy justice can amend the ills of local content policy is through the adaptation of restorative justice in further local content implementation. Restorative justice emanates from the presence of harm being experienced by victims or societies.[24] It is rehabilitative. Thus it extends the purpose of

[24]Heffron, R.J. and McCauley, D., 2017. The Concept of Energy Justice Across the Disciplines. *Energy Policy*, 105, pp. 658–667.

the law from being punitive to being more inclusive and rehabilitating the offenders back into society. Sullivan and Tifft state that "restorative justice makes society think about how to respond to injustices that have occurred and also in defining what injustices society give attention to in the first place".[25]

Though restorative justice has mainly been developed in criminal law, in this book restorative justice is essential in finding ways to respond to the harm being done by the adoption and implementation of inadequate local content policies. Restorative justice is the mediator on how policies can avoid political patronage, rent-seeking and fronting. These policies can facilitate economic diversification without favouring political and ethnic elites. Ovadia explains the effects of local content policies in Nigeria by stating that "there are some positive signs that the policies will have an impact across more significant segments of the Nigerian population, the benefits are stacked in favour of the elite". Nigerian content represents a significant alteration in the political and economic structure of Nigeria's oil and gas industry.

In comparison, the effect of Nigerian content policies on the industry will play out in a domestic and geopolitical context. The Nigerian petroleum elite has made an organised and concerted effort to design and implement policies to maximise their benefit in terms of power and wealth. Ovadia states that "If a more capitalistic economy emerges from these policies, it will be because those elites have made a conscious effort to embrace new strategies of accumulation based on the production of value for the Nigerian economy".[26] In responding to such harms, restorative justice will force policymakers to take into account principles of energy justice while designing, adopting and implementing local content policies in throughout their energy lifecycle. Heffron and McCauley state that "in some cases, these costs of restoration would be prohibitive and consequently that energy activity would cease or not be proposed".[27] In

[25]Sullivan, D. and Tifft, L., 2006. Introduction: The Healing Dimension of Restorative Justice—A One-World Body.*Handbook of Restorative Justice* (pp. 1–16).

[26]Ovadia, J.S., 2013. The Making of Oil-Backed Local Capitalism in Nigeria. *New Political Economy*, 18(2), pp. 258–283.

[27]Heffron, R.J. and McCauley, D., 2017. The Concept of Energy Justice Across the Disciplines. *Energy Policy*, 105, pp. 658–667.

the application of the principle to local content policy, for this book, the national and community content policies will not be discussed as a channel for efficient integration of the local population into the energy value chain. The two policies have a considerable influence of the political elite since the national policymakers are designing the policy, making the policy vulnerable to patronage and corruption, making the policy fall short on distributional and procedural justice.

Due to the three tents of energy justice, they are not resonating with community content and national content policy. It is essential to identify how restorative can apply to local content policies. Local content policies are vital for the development of local workforce and industries suing oil and gas, through the development of forwarding and backward linkages. In this book, restorative justice is essential in developing a regional content policy and framework. The implementation of the regional content policy is done through regional cooperation among different countries. The interests of oil and gas being dealt with at a regional level can only amount to fruition if the interests of justice have been taken into account while adopting and implementing a regional content framework.

3.5 Levels of Energy Law and Restorative Justice

It should be noted that energy law is derived from three levels of law, namely international, national and local law.[28] In the international segment, countries enter into international agreements. These agreements or international instruments become the driving force behind the energy sector among the party member states or the States that ratified the agreement. The segment is the national level of energy law, the country's policymakers make decisions on behalf of the public with regard to energy law dependent on the aim of the government, the availability of finances and technology as well as the societal preferences at

[28] Heffron, R.J. and Talus, K., 2016. The Development of Energy Law in the 21st Century: A Paradigm Shift? *The Journal of World Energy Law & Business*, 9(3), pp. 189–202.

the time. Lastly is the local level of energy law, this is the smallest unit of energy law, and it is the level of energy law that involves the host community of oil and gas. In this level of energy law, individuals are personally affected by the presence of oil and gas in their community.[29] The level of energy law applicable is the international energy law in terms of applying restorative justice principles into creating a regional content framework, whereby two or more different countries can have an agreement towards creating energy policy. The restorative justice will avail a transparent decision-making process since the eight principles of energy justice will provide the framework under which a regional content policy is created thereof.

It should be noted that restorative justice has previously been instrumental in different phenomena such as Environmental Impact Assessments, Social licence and Energy Financial Reserve Obligation[30]. Through restorative justice the regional content policy will ensure (1) the **availability** of suppliers who can pool functional technical capacity to the energy industry, (2) the suppliers will avail goods and services that are **affordable** at an internationally competitive price, quality and readily available when needed, (3) the process of adopting and implementing a regional content policy will follow **due process** to avoid the loopholes found in the conventional local content policies, namely the national and community content, (4) the principles of **good governance** will be underpinned to minimise corruption and improve accountability among the energy industry stakeholders, (5) The regional content policy will be **sustainable** to adopt skills and technology that will be applicable even during the energy transition, (6) the regional content policy will avail everyone within the region that adopted the framework right to access a stake in the oil and gas through **intra-generational equity** as a principle of distributive justice, (7) the regional content policy also holds a moral responsibility to ensure that the exhaustion of the oil and gas within the region by the local must ensure that today's children and future generations inherit a global environment at least no worse than the one we

[29] Heffron, R.J. and Talus, K., 2016. The Development of Energy Law in the 21st Century: A Paradigm Shift? *The Journal of World Energy Law & Business*, 9(3), pp. 189–202.

[30] Ibid.

received from our predecessor through principles of **inter-generational equity** and (8) the regional content policy must take into account the **responsibility** of the party States in ensuring that they create a competitive advantage for the local population while minimising the production of negative externalities.[31]

3.6 Summary

Local content policies are pivotal in natural resource governance. But the demands of this industry are expensive and capital intensive, with a need for highly sophisticated skills that developing oil and gas producers do not possess. This maximises the enclave nature of the oil and gas industry, international oil and gas companies fly in with their backward and forward linkages intact minimising the depth and breadth of the oil and gas industry to other sectors of the economy. Inadvertently, hindering economic growth leading to a resource curse. This chapter addresses these issues from a conceptual basis using critical elements of justice-distributive, procedural, restorative, recognition and cosmopolitan. This interdisciplinary approach towards analysing local content policies address problems on social license to operate, cluster formations, skills pooling in cross-border actions and cooperation, accumulation of capital for the sole purpose of cracking through the enclave nature of the oil and gas industry in Sub-Saharan Africa.

There is a need for society to ensure a more equitable distribution of resources given the current state of the world due to COVID-19, energy transition and the oil price shocks of 2020. We are considering the timeline of the oil and gas life cycle and the gradual shift of investments from fossil fuels to cleaner sources of energy due to the energy transition. It is scary that the skills acquired by developing oil and gas producers in Sub-Saharan Africa might be redundant in a few years. Energy justice advocates for inter-generational and intra-generational equity. There is hope that local content can be designed, adopted and implemented in

[31] Sovacool, B.K. and Dworkin, M.H., 2015. Energy Justice: Conceptual Insights and Practical Applications. *Applied Energy*, 142, pp. 435–444.

light of energy justice. We can achieve a fair distribution of the benefits of our natural resource

References

Fraser, N., 2009. Social Justice in the Age of Identity Politics. *Geographic Thought: A Praxis Perspective* (pp. 72–91).

Heffron, R.J., 2014. *Energy Law: An Introduction.* Springer.

Heffron, R.J. and McCauley, D., 2017. The Concept of Energy Justice Across the Disciplines. *Energy Policy*, 105, pp. 658–667.

Heffron, R.J. and Talus, K., 2016. The Development of Energy Law in the 21st Century: A Paradigm Shift? *The Journal of World Energy Law & Business*, 9(3), pp. 189–202.

Heffron, R.J., Rønne, A., Tomain, J.P., Bradbrook, A. and Talus, K., 2018. A Treatise for Energy Law. *The Journal of World Energy Law & Business*, 11(1), pp. 34–48.

Jenkins, K., McCauley, D. and Forman, A., 2017. Energy Justice: A Policy Approach. *Energy Policy*, 105, pp. 631–634.

Jenkins, K., McCauley, D., Heffron, R., Stephan, H. and Rehner, R., 2016. Energy Justice: A Conceptual Review. *Energy Research & Social Science*, 11, pp. 174–182.

McCauley, D.A., Heffron, R.J., Stephan, H. and Jenkins, K., 2013. Advancing Energy Justice: The Triumvirate of Tenets. *International Energy Law Review*, 32(3), pp. 107–110.

Ovadia, J.S., 2013. The Making of Oil-Backed Local Capitalism in Nigeria. *New Political Economy*, 18(2), pp. 258–283.

Schlosberg, D., 2003. The Justice of Environmental Justice: Reconciling Equity, Recognition, and Participation in a Political Movement. *Moral and Political Reasoning in Environmental Practice* (Vol. 77, p.106).

Sovacool, B.K. and Dworkin, M.H., 2015. Energy Justice: Conceptual Insights and Practical Applications. *Applied Energy*, 142, pp. 435–444.

Sovacool, B.K., Heffron, R.J., McCauley, D. and Goldthau, A., 2016. Energy Decisions Reframed as Justice and Ethical Concerns. *Nature Energy*, 1(5), p. 16024.

Sullivan, D. and Tifft, L., 2006. Introduction: The Healing Dimension of Restorative Justice—A One-World Body. *Handbook of Restorative Justice* (pp. 1–16).

Todd, H. and Zografos, C., 2005. Justice for the Environment: Developing a Set of Indicators of Environmental Justice for Scotland. *Environmental Values*, 14(4), pp. 483–501.

Valerie Marcel and others, 2016. A Local Content Decision Tree for Emerging Producers (Energy, Environment and Resources Department Chatham House: The Royal Institute of International Affairs).

Walker, G., 2012. *Environmental Justice: Concepts, Evidence and Politics*. Routledge.

4

In-depth Synopsis of Local Content Literature

4.1 Background of Local Content Policies

The first notable work of local content was by an American economist Gene M. Grossman published in 1981. He evaluated the impact of local content protection measures on the market structures and the production of intermediate goods.[1] Grossman believed that a given percentage of the products made by the local embodied in the final product. Grossman's analysis of local content assumed that the intermediate and final goods are produced in a perfectly competitive market.[2] However, Schott and Hufbauer stated that "local content policies create a production distortion and diminish incentives for technical efficiency, though local content policies can be the second-best policy instrument because they avoid the consumption distortion created by a tariff".[3]

[1] Grossman, G.M., 1981. The Theory of Domestic Content Protection and Content Preference. *The Quarterly Journal of Economics*, 96(4), pp. 583–603.

[2] Ibid.

[3] Hufbauer, G.C., Schott, J.J. and Cimino-Isaacs, C., 2013. *Local Content Requirements: A Global Problem* (Vol. 102). Columbia University Press.

R. S. Muhongo, *Energy Justice*, Energy, Climate and the Environment,
https://doi.org/10.1007/978-3-030-61338-9_4

Several industries did implement these protectionist measures. The Canadian-American Automotive agreement in 1965, had protectionist content elements used in Canada to protect the automotive parts industry from US competition.[4] Other countries, such as New Zealand and Australia, adopted content protection schemes. Other notable economists such as Findlay and Welisz held that this type of protection is popular in industries where the final product are assembled from different parts, which may be imported or manufactured domestically.[5] This has been experienced in Australia for automobiles, tractors, industrial machinery, petrochemicals, and various consumer nondurables. However, in developing countries, domestic content rules are weak for newly established industries they can't locally produce any components or parts that are needed in the local industries. Thus the protection needs to gradually be tightened to promote local production of the given parts and components.[6] Before Grossman, theoretical analysts such as Wonnacott and Wonnacott, Corden. They explained that content protection is defined in physical terms, the assumption is purely based on competitive market structure, and intermediates are used only in proportions fixed to output.[7]

This entails that content protection will not be efficient in weak economies. Furthermore, the service industry is not protected in this sense since content protection is only defined in physical terms. Local content policies have been applied in different countries, different industries and in different contexts to meet the different economic and social-political requirements needed by the policymaker. This chapter provides a full literature review on local content policies and other supervening factors that affect the adoption and implementation of local content policies as well as the metrics used to assess the effectiveness of these policies. The literature review further searches for the appropriate

[4]Lloyd, P.J., 1973. *Non-tariff Distortions of Australian Trade*. Canberra: Australian National University Press.

[5]Wellisz, S. and Findlay, R., 1988. The State and the Invisible Hand. *The World Bank Research Observer*, 3(1), pp. 59–80.

[6]Wellisz, S. and Findlay, R., 1988. The State and the Invisible Hand. *The World Bank Research Observer*, 3(1), pp. 59–80.

[7]Corden, W.M., 1985. *Protection, Growth and Trade: Essays in International Economics*. Oxford: Blackwell.

definition of local content that is not subjected to a particular jurisdiction. The chapter also provides a critical review of the other theoretical constructs other than energy justice such as regionalism and localism and their application to local content policies. The chapter portrays local content policies in an interdisciplinary perspective other than just a policy that is limited to economic and political connotations. It also entails the different objectives of local content policies across different jurisdictions. The chapter also provides the alternatives of local content as well as the application of local content policies in the oil and gas sector. The chapter elaborates on linkages and factors that affect the adoption of linkages in different economies. Lastly, an assessment is made on how energy justice identifies the gap in local content implementation.

4.2 Searching for a Definition of LCP

Cimino, Huffbauer and Schott define local content policies as a "deliberate preference" or a "first consideration" to the country's nationals in matters of employment and training and the procurement of goods and services.[8] Several countries have adopted the definition of local content policy in ways similar to Cimino, Huffbauer and Schott's definition of "deliberate preference" or a "first consideration". The form of placing a deliberate preference has been mostly adopted by governments, such as Nigeria, Ghana Angola and Tanzania. In Norway, local content has no definition; the policymakers did not limit the application of the policy through a specific definition. In Brazil local content is defined as the proportion between the value of the goods produced and services rendered in Brazil per the terms of the Model Contract 2015 and the total value of the goods used and services provided for this purpose.[9]

The Nigerian government, the Nigerian National Petroleum Corporation (NNPC) defines local content as "the quantum composite value added or created in the Nigerian economy through the utilization of

[8]Cimino, Cathleen, Hufbauer, Gary Clyde and Schott, Jeffrey J., A Proposed Code to Discipline Local Content Requirements. Peterson Institute for International Economics Policy Brief No PB14–6, February 2014, p. 1.

[9]Model Contract 2015, Art. 1.3.15.

Nigerian human and material resources for the provision of goods and services to the petroleum industry".[10] Ghana's Petroleum (Local Content and Local Participation) Regulations, 2013 defines local content as "the quantum or percentage of locally produced materials, personnel, financing, goods and services rendered in the petroleum industry value chain and which can be measured in monetary terms".[11] The two definitions of Ghana and Nigeria emphasise the utilising local companies, local people as well as having an input of the local product (such as raw materials) in the creation of the final product. The definition also advocates for backward and forward linkages. This is because the definitions entail the local participation in the whole energy value chain from consumption to waste management. The two definitions have adopted the so-called national content policy approach. This approach of local content policies applies to the whole nation. It does not place special consideration on the host communities that face the ramifications of the energy industry within their vicinity.

This has led to NYMBISM (Not in my backyard syndrome), of these oil and gas. Hence the national content approach falls short in distributive justice. Since most of the benefits are fundamentally not shared by the host community but rather the negative consequences of having such oil and gas. This has been seen in Niger Delta where oil spills, environmental pollution have been prevalent, unlike the non-oil-rich communities. Tanzania in its Local Content Policy for the Oil and Gas Industry drafted in 2014. Defines local content as "the added value brought to the country in the activities of the oil and gas industry in the United Republic of Tanzania through the participation and development of local Tanzanians and local businesses through national labour, technology, goods, services, capital and research capability".[12] Tanzania further defines local content as the added value brought to a host nation

[10]Nwapi, C., 2015. Defining the "Local" in Local Content Requirements in the Oil and Gas and Mining Sectors in Developing Countries. *Law and Development Review*, 8(1), pp. 187–216.

[11]Petroleum (Local Content and Local Participation) Regulations, 2013, LI 2204, Reg 49, available at: http://www.reportingoilandgas.org/wp-content/uploads/PETROLEUMLOCAL-CONTENT-AND-LOCAL-PARTICIPATION-REGULATIONS2013.pdf.

[12]United Republic of Tanzania, Local Content Policy (2014).

(and regional and host energy resource communities in that country) through the activities of the oil and gas industry".[13]

The terms used such as "regional and local areas" suggests that there is specific recognition given to the host communities of these oil and gas within the country. The emphasis placed by the policy on energy resource host communities is coherent to justice as recognition. Commonly, oil and gas are found in remote areas where these areas are isolated before the discoveries of oil and gas, such as the gas-rich Mtwara region in Tanzania, the oil-rich Turkana county of Kenya and the oil-rich Albertine region of Uganda. The discoveries create a gold rush that can only be addressed once there are adoption and implementation of a community content approach.

The mention of host communities triggers what policy analyst Micheal Warner has argued for, a phenomenon known as "community content".[14] Community content holds that the host community must be involved (given first preference) in the energy lifecycle. Thus there must be a strategic deployment of local participation and local capability development opportunities in the energy industry. Warner states that "community content relates to realising a competitive advantage for the oil company in the eyes of both the local population and the country's guardians of economic policy".[15] In the energy, justice community content is regarded as restorative justice. Warner mentions that community content programme is a "merit good" that targets those negatively impaired.

The oil and gas host communities are usually the communities that suffer the most due to the extractive industry. Though the foundations of restorative justice are found in criminal justice, in energy law, the principles of restorative justice aim to repair the harm done to the local

[13]Ibid.

[14]Warner, M., 2007. Community Content: The Interface of Community Investment Programmes with Local Content Practices in the Oil and Gas Development Sector. Overseas Development Institute (ODI) Briefing Note (9).

[15]Warner, M., 2007. Community Content: The Interface of Community Investment Programmes with Local Content Practices in the Oil and Gas Development Sector. Overseas Development Institute (ODI) Briefing Note (9).

victims of the oil and gas.[16] That is the similarity between community content and energy justice; they both aim to repair the damage done to the host community by the mere presence of the oil and gas within their vicinity. Both community and national content policies exclude foreigners, while the community content excludes foreigners and non-community members. The national content policy excludes foreigners from different nationals. However, Warner believes that community content should be distinct from the national content policy. Community content cannot be materialised as most host communities do not possess the capability. They are technologically incapable, with low capital, low industrial base, and poor infrastructure. There usually is insufficient technical skills to service the energy industry.

Nwapi thinks that community content and national content policy can be implemented at the same time; it is well understood that there will be no synergy with justice in recognition. Labelle states that justice in recognition "provides a cultural and political representation of groups with distinguishing features such as social, ethnic and gender differences; freedom distortion, devaluation, degradation and physical threats by other groups".[17] Hence to advocate for both community and national content in the same energy policy cannot be realised in host communities. This will have a detrimental supply of benefits and ills of the effects of oil and gas. As the host communities will suffer from the ills and the other non-host communities will mainly benefit from the resources.

Community and national content policies do not have a holistic approach towards local integration of the energy industry. A critical challenge for developing energy resource-rich countries is adopting and implementing local content policies that reflect their peculiar energy industry needs. The challenges that most oil and gas-rich countries face individually. A regional policy approach towards local content would be more efficient. Lalbahadur argues that a plausible way would be through

[16]Heffron, R.J. and McCauley, D., 2017. The Concept of Energy Justice Across the Disciplines. *Energy Policy*, 105, pp. 658–667.

[17]LaBelle, M.C., 2017. In Pursuit of Energy Justice. *Energy Policy*, 107, pp. 615–620.

"a more strongly focused African strategy to promote the technology and skills insufficiency in Africa".[18]

Furthermore, The Oxford Institute for Energy Studies argues that countries with small reserves (600 million barrels) but play a pivotal role in their region such as Kenya.[19] They might have a more significant role if they acted as an energy industry regional hub in the East African region. This can be achieved by Kenya sharing its facilities with its neighbours and being more inclusive through a regional content policy since most developing countries cannot enter into the sophisticated high capital-intensive energy industry. They were using the comparative advantage to mobilise the capability that can enable the local energy industry participants to meet the demands of the energy industry. Nwapi further states that "while the idea behind LCPs is to promote local participation in FDI, restrictions to the employment of foreign labour may hurt a country's economic growth by restricting the cross-border exchange of information and technology transfer and denying local professionals learning opportunities that the presence of foreign professionals provides".[20] Most of the above definitions do not show that local content provisions require government intervention for such as a policy to qualify as a local content policy. This is because corporate social responsibility may be undertaken as local content policies adopted by the multinational corporations. Since they may entail the same provisions such as training of the local labour force, technological transfer, entrepreneurship development programmes, etc. Conclusively on the search for a definition of what local content is and what it should entail. The book provides a definition that will be adopted throughout: Local content is defined as, the legislative, administrative and judicial government intervention in the oil and gas industry by ensuring local-preference in the provisions of goods and services as well as in the

[18]Lalbahadur, Aditi, A Regional Approach to National Development Priorities in Mozambique? 20 March 2013, South African Institute of International Affairs, online: https://saiia.org.za/research/a-regional-approach-to-national-development-priorities-in-mozambique/.

[19]Patey, Luke, Kenya: An African Oil Upstart in Transition. Oxford Institute for Energy Studies Paper: WPM 53, p. 23, online: http://www.oxfordenergy.org/wpcms/wp-content/uploads/2014/10/WPM-53.pdf.

[20]Nwapi, C., 2015. Defining the "Local" in Local Content Requirements in the Oil and Gas and Mining Sectors in Developing Countries. *Law and Development Review*, 8(1), pp. 187–216.

authorization of oil and gas. Goods and services may include technology, R&D, labour, rig provisions, engineering and design services, well-heads, completion equipment, laboratory services, supply chain services, vehicle fleet management, recruiting, HR training and health, financial and risk services, construction services, site support consumables, lifting services, etc.

4.3 Critical Review of the Theoretical Literature on Local Content Policies

4.3.1 Localism Theory

Localism, as a theory places its focus on local empowerment. The principles are found in the role of local governments in promoting the efficient delivery of public goods and services.[21] The fact that more efficiency is expected when there is local autonomy, this creates specificity when it comes to tailoring public policy and decision making, especially in energy policy concerns. Host communities affected by the presence of oil and gas need to take a consultative approach and have an inclusive mechanism of those affected by it the most. Local governments have direct access to the most impacted people of these oil and gas. The localism theory approach tends to make an opinion that national governments cannot well devise policies that are peculiar to the oil and gas host communities. At the local government level, however, there is a greater chance that the voices of local people can influence public decisions.[22] In light of the local content policies through the localism theory emphases that the development of local content policies must be tailored under the

[21] Briffault, Richard, Localism and Regionalism, Columbia Law School Public Law and Legal Theory Working Paper No 1, 11 August 1999, p. 18.

[22] Briffault, Richard, Localism and Regionalism, Columbia Law School Public Law and Legal Theory Working Paper No 1, 11 August 1999, p. 18.

local autonomy of the host communities that have their distinct values and identities as well as historical experiences.[23]

Through localism, the development of a community content approach can also help solve the problems brought about by oil and gas in remote areas. Hence governments can efficiently address the host community's challenges through achieving "political harmony".[24] Localism can be utilised to enhance the inclusiveness of the host communities like the Niger Delta, Turkana, Mtwara and Albertine region community. These energy host communities are usually excluded from the benefits of the oil and gas despite hosting them.

4.3.2 Regionalism Theory

Regionalism advocates that a region a fundamental economic, social, and ecological unit.[25] Such unit as defined by the theory of regionalism entails that people may live in one area and have their dealings or business activities in another area. Furthermore, this would entail that there is movement between one or more municipalities to fulfil their socio-economic needs. The regionalism theory brings about regional integration that encourages there to be cooperation due to technical, functional purposes. The process will gradually spill-over within and across sectors. According to the neo-functionalists such as Haas argue that the spill-over effect would result in increased political integration, supra-nationality, and a re-definition of group identity around what is known as the regional unit.[26]

Regionalism has been used as a tool to avail solutions that may not be found within the national boundaries of a given nation. The pooled efforts from different actors and stakeholders such as the state, market and civil society within a given geographical area to increase economic

[23]Kessler, Anke S., Hansen, Nico A. and Lessmann, Christian, 2011. Interregional Redistribution and Mobility in Federations: A Positive Approach. *Review of Economic Studies*, 78, p. 1345.

[24]Warner, Michael, 2011. *Local Content in Procurement: Creating Local Jobs and Competitive Domestic Industries in Supply Chains*. Sheffield, UK: Greenleaf Publishing.

[25]Briffault, R., 2000. Localism and Regionalism. *Buffalo Law Review*, 48, p. 1.

[26]Haas, Ernst B., 1964. *Beyond the Nation-State*. Stanford: Stanford University Press.

development of the region and improve the region's socio-economic position.[27] It can bolster a region's position by creating novel mechanisms that would meet the demands and needs of the high capital-intensive energy industry within a given region. The regional content policy would not entail the creation of new regional institutions. But through the existing local stakeholders operating in the energy industry. The different collaborations between local stakeholders of the same region. Can create a national competitive advantage through innovation and the pooling of capacity to integrate the energy industry value chain. Porter states that "differences in national values, culture, economic structures, institutions and histories all contribute to competitiveness success".[28] Thus for local companies to compete and have an advantage against the multinational corporations in their energy industry, this can only be achieved through regionalist measures such as clustering, pooling of resources, free labour and capital movement, etc.

4.3.3 Local Content Policies as Interdisciplinary Perspectives

Local content policies are interdisciplinary. They are policies that are used in the legal, economic and political science disciplines in their knowledge exhaustion and application. A local content policy is found within the legal framework of a country. The provisions of these policies are found in primary and secondary legislation as well as in negotiated agreements.[29] These agreements and legislation dictate how local energy participants will participate in the energy industry. These policies found their origin in economics as a domestic content measure. Whereby economists such as Grossman argued, "A content protection scheme requires that a given percentage of domestic value-added or

---

[27] Hettne, B. and Söderbaum, F., 1998. The New Regionalism Approach. *Politeia*, 17(3), pp. 6–21.

[28] Grant, R.M., 1991. Porter's 'Competitive Advantage of Nations': An Assessment. *Strategic Management Journal*, 12(7), pp. 535–548.

[29] Tordo, S., Warner, M., Manzano, O. and Anouti, Y., 2013. *Local Content Policies in the Oil and Gas Sector*. The World Bank.

domestic components be embodied in a specified final product".[30] But later Grossman himself concluded that "the extent of protection or preference is not readily predictable, content protection and content preference may fail to attain the non-economic objectives of the policy-maker".[31] Local content policies are also found in the realm of political scientists. Ovadia has stated that local content policies are used by the elite to consolidate power in the development of a petro-state such as Angola and Nigeria.[32] Hence local content policies are said to have a dual nature whereby one hand facilitates local integration in the energy value chain. But the legitimacy of the said integration inputting to question since it serves the bidding of the elite structure, especially in developing energy resource-rich countries.

4.3.4 Conclusion

The two competing theories of regionalism and localism both offer viable arguments in their application of local content policies. As much as the localism theory advocating for local empowerment, the regionalism theory brings about a more comprehensive approach that encompasses broader scales of production in the adoption and implementation of local content policies. Through regional integration alone is not a viable approach to meeting the demands of the oil and gas industry as well as ensure that the local communities gain economic development from their resources. Energy justice as a theory brings a resolutive approach since energy justice calls for recognition of all marginalised communities; this shows that the objectives of the localism theory will be met. Energy justice further entails that there should be a "just" application when adopting and implementing such policies to avoid any harm to be done to the different stakeholders. This aspect of energy justice accepts the use of regional integration as a tool, through regional integration all

---

[30] Grossman, G.M., 1981. The Theory of Domestic Content Protection and Content Preference. *The Quarterly Journal of Economics*, 96(4), pp. 583–603.

[31] Ibid.

[32] Ovadia, J.S., 2012. The Dual Nature of Local Content in Angola's Oil and Gas Industry: Development vs. Elite Accumulation. *Journal of Contemporary African Studies*, 30(3), pp. 395–417.

other tenets of energy justice can be met to ensure the "just" adoption of local content policies thereof.

4.4 Objectives of Local Content Policies

Local content policies are implemented based on a given objective. The oil and gas industry is a very lucrative industry that creates expectations within the general population. Most governments intervene in the oil and gas industry to meet the demands of the general population such as creating jobs, creating new markets for the local goods and services, as well as bring about new technology. Since the book cannot summarise all provisions of numerous local content policies, the objectives of these policies have been pooled in these five objectives, namely the provisions for local employment, local goods and services, provisions for training of the local workforce, technology transfer and local business development.

4.4.1 Provision of Local Employment

A local content policy advocates for the hiring of local nationals. The employment provision can entail unskilled, skilled and semi-skilled labour. In Angola, a minimum of 70% of Angolan citizens must be hired in the workforce by headcount.[33]

4.4.2 Provision of Local Goods and Services

Local content policies advocate for local goods and services. These requirements are specifically targeting a minimum percentage for the provision of goods and services. Requirements include:

- A given proportion for the provision of goods and services.[34]

[33] Law 10/79 mandating Sonangol to Employ and Train Nationals.

[34] Decree 13/10 advocates for goods and services not requiring heavy capital investment and with a basic, medium or higher level of non-specialised know-how are reserved for Angolan companies. Foreign companies may take part in the initiative of Angolan companies.

- A given preferential treatment policy.[35]
- The joint-venture requirement between foreign and local companies.

4.4.3 Provision for Training of Local Workforce

International oil companies are to train the local workforce. Developing the local workforce is essential for the inclusion of the local workforce in the oil and gas value chain. The Indonesian local content policy designed to promote nationals' skills development. Companies investing in Indonesia must have training programmes to achieve local target levels.[36]

4.4.4 Provision for Technological Transfer

Local companies may not have the technological capabilities to meet the demands of the oil and gas industry. In Brazil, the operators must invest at least 1% of their gross revenue in technological research of oil and gas development.[37]

4.4.5 Provision for Local Business Development

Local content policies are also adopted to develop the business climate and foster international and local investment. This helps create a strategic approach to increase the value-added within a given country. In Kazakhstan, the local content policy adopted a price discrimination mechanism to favour local companies. Furthermore, other measures such as interest-free loans, advanced payment for local firms and joint ventures (between a foreign and a local firm), support in technological transfer

[35]The Angolanization policy also advocates for national private or state-owned suppliers are preferred over the suppliers if the price offered is no more than 10% higher than that of foreign suppliers.

[36]Presidential Regulation No. 54/2010 and The Oil and Gas Law No. 22/2001 Segregating Regulatory Roles and Commercial Operations.

[37]The Brazilian Petroleum Law Number 9478/97.

initiatives.[38] Though local content policies have the above objectives, scholars such as Grossman, Cimino, Hufbauer and Schott have given reasons to believe that local content policies are not favourable policies. Since the protection afforded by local content policies is highly variable. They also hold the opinion that local content policies are not transparent.[39] This makes the policy vulnerable to corruption.

4.5 The Development of Local Content Metrics

Local content metrics are used to assess the delivery of local content requirements and local content development. The metrics assess the extent that the local people or local suppliers are capable of meeting the demands of the energy industry. These metrics can be through the number of local people in the energy industry counted by headcount within a given workforce. Can also be the expenditure spent on local suppliers and or the participation of locally owned companies in contract execution.[40] Local content share can be grown over time, for example, by building skills in the national workforce through training and education, or investing in developing the capabilities and competitiveness of domestic supply chains. Angolan nationals form local content implementation countries such as Angola whereby Decree 5/95 prevents companies from employing non-resident foreigner workers unless at least 70% of the workforce.[41]

Furthermore, policy metrics can be broken down into different subcategories such as professional positions or non-professional positions or some other disaggregation of employment positions in this regard.

[38] Law 291-IV/2010 on Subsoil and Subsoil users.

[39] Hufbauer, G.C., Schott, J.J. and Cimino-Isaacs, C., 2013. *Local Content Requirements: A Global Problem* (Vol. 102). Columbia University Press.

[40] Tordo, S., Warner, M., Manzano, O. and Anouti, Y., 2013. *Local Content Policies in the Oil and Gas Sector*. The World Bank.

[41] Employment of Foreign Citizens Non-Resident, Decree 5/95, Ministry of the Interior, Republic of Angola, http://www.sme.ao/attachments/article/209/Decree%20No.%20595%20of%20April%207%20%20Employment%20of%20Foreign%20Citizens%20Non-resident.pdf.

The Production Sharing Agreement for the Shah Deniz Prospective Area in the Azerbaijan Sector of the Caspian states the following metrics "Before the commencement of development: professionals 30–50 per cent; non-professionals 70 per cent, upon commencement of petroleum production: professionals 70 per cent; non-professionals 85 per cent, five years after commencement of petroleum production: professionals 90 per cent; non-professionals 95 per cent".[42] The metrics for measuring local content in supply chains are divided into two approaches within the expenditure of goods and services. Warner states that "The first seeks to classify expenditure against some definition of a local, national or domestic business entity, that is, the producer or provider of the good or service. The second classifies expenditure by some measure of whether the relevant service or good is deemed to be local and, for goods only, whether the value has been added in-country".[43] The International Financial Corporation (IFC) further holds that the definition of local can be reached through one or more of the following criteria. Regulatory, geographic, ownership, size of the enterprise and the under-represented groups.[44]

The IFC further hold that the legal framework of a country clearly defines what is regarded as local. The guide describes the geographic region criteria as the communities that are directly impacted by operations, and these places have high expectations for jobs and other opportunities in the oil and gas industry.[45] Local ownership can be supplier owned by foreign capital but located in country X, supplier partly owned by locals with a foreign majority stake, supplier partly owned by locals with local majority stake and supplier wholly owned by

[42]Agreement on the Exploration, Development and Production Sharing for the Shah Deniz Prospective Area in the Azerbaijan Sector of the Caspian Sea, 1996, http://www.bp.com/liveassets/bp_internet/bp_caspian/bp_caspian_en/STAGING/local_assets/downloads_pdfs/pq/SD_PSA.pdf.

[43]Warner, M., 2017. *Local Content in Procurement: Creating Local Jobs and Competitive Domestic Industries in Supply Chains*. Routledge.

[44]A Guide to Getting Started in Local Procurement: For Companies Seeking the Benefits of Linkages with Local SMEs. International Financial Corporation. http://www.ifc.org/wps/wcm/connect/03e40880488553ccb09cf26a6515bb18/IFC_LPPGuide_PDF20110708.pdf?MOD=AJPERES.

[45]Ibid.

locals.[46] Tordo has a combination of criteria to measure expenditure on local suppliers: "The address is given in vendor registration information, the address on a purchase order or invoice, the geographic location of the service being provided or the production of the good being supplied, the share of equity owned by national citizens (for example, greater than 25 per cent, 50 per cent or 100 per cent), whether the supplier is locally incorporated, whether the supplier is tax registered in the country of operation, including for withholding tax purposes, whether the supplier employs more than a specified percentage of nationals"[47]. Criteria for regional or community suppliers as "The regional/local address in vendor registration, the regional/local address on invoice, suppliers and contractors who source the majority of materials or labour from the province, district, or local communities located closest to the operation".[48] The measure of the development of local suppliers is also in two main categories. These are the following the input metrics and the output metrics.[49]

4.5.1 Input Metrics for Measuring Local Supplier Development

Input metrics are essentially alternative indicators for performance assessment. This assessment is more value-based. It measures the direct proportionality between the costs of vocational and or educational training of the local employees to the value of their future earnings in the energy sector.[50]

[46]Warner, M., 2017. *Local Content in Procurement: Creating Local Jobs and Competitive Domestic Industries in Supply Chains*. Routledge.

[47]Tordo, S., Warner, M., Manzano, O. and Anouti, Y., 2013. *Local Content Policies in the Oil and Gas Sector*. The World Bank.

[48]Warner, M., 2010. Unpacking Local Content Metrics and Measurement Dr Michael Warner, November PDF. [online] Docplayer.net., available at: https://docplayer.net/48696189-Unpacking-local-content-metrics-and-measurement-dr-michael-warner-november-2010.html. Accessed 10 September 2019.

[49]Tordo, S., Warner, M., Manzano, O. and Anouti, Y., 2013. *Local Content Policies in the Oil and Gas Sector*. The World Bank.

[50]Warner, M., 2010. Unpacking Local Content Metrics and Measurement Dr Michael Warner, November PDF. [online] Docplayer.net., available at: https://docplayer.net/48696189-Unpacking-local-content-metrics-and-measurement-dr-michael-warner-november-2010.html. Accessed 10 September 2019.

4.5.2 Output Metrics for Measuring Supplier Development

Output metrics are more viable than input metrics, they are robust, and they measure the actual improvements of local suppliers over a given period. Output metrics measure the local firm capability to compete with its counterparts in the same energy industry.[51] They create measurable parameters such as supplier costs, delivery times, labour productivity, duality, volume capacity and HSE performance.

4.6 Alternatives to Local Content Policies

The alternatives to local content policies that can meet the objectives of local content policies.

4.6.1 Implementing Performance Requirements

Sauve defines performance requirements as the "stipulations, imposed on investors, requiring them to meet certain specified goals concerning other operations in the host country".[52] Performance requirements must be aligned with other policies such as trade policy, depletion policy, procurement policy, and industrial policy to meet numerous policy objectives.[53] Sauve further states that performance requirements have the following objectives "strengthening a country's industrial base and increasing, domestic value-added, generating expanded employment opportunities, promoting vertical linkages between lead (foreign) firms and (local) suppliers, increasing exports, balancing trade, promoting regional development and spatial gaps in development levels, fostering

[51] Ibid.

[52] Sauvé, P., 2016. Life Beyond Local Content: Exploring Alternative Measures of Industry Support in the Context of WTO Accession. *Journal of International Trade*, 1, pp. 1–28.

[53] Moran, T.H., Graham, E.M. and Blomström, M. (Eds.), 2005. *Does Foreign Direct Investment Promote Development?* Peterson Institute.

technology transfers, avoiding restrictive business practices, generating and distributing economic rents; and pursuing various non-economic objectives, such as political independence, the promotion of minority rights or the (re)distribution of political power".[54] This alternative if adopted can meet the objectives of local content policies.

4.6.2 Creating a Local Business Friendly Environment

Researchers such as Kolstad and Kinyondo state that "a better way to create jobs and stimulate investment is to upgrade the national environment for doing business".[55] Foreign investors are concerned with economic and political investment climate of a host country. The most accessible mechanism to determine the risk associated with foreign direct investment is through indices from recognised institutions such as the World Bank. The World Bank has a Doing Business Report. The 2019 Doing Business Report in 2019 had the following indicators "starting a business, dealing with day to day operations, accessing finance, getting a location".[56] If a country has a good ranking in the measured indices, it will automatically attract foreign direct investment. Foreign direct investment adopts a corporate social responsibility as a corporate practice to gain a social license to operate. The presence of direct social responsibility in itself meets the local content policy objectives.

4.6.3 Adopting Strict Corporate Social Responsibility

Votaw describes corporate social responsibility as "something but not always the same thing to everybody. To some, it conveys the idea of legal responsibility or liability; or a socially responsible behaviour in the

[54] Sauvé, P., 2016. Life Beyond Local Content: Exploring Alternative Measures of Industry Support in the Context of WTO Accession. *Journal of International Trade*, 1, pp. 1–28.

[55] Kolstad, I. and Kinyondo, A., 2017. Alternatives to Local Content Requirements in Resource-Rich Countries. *Oxford Development Studies*, 45(4), pp. 409–423.

[56] Doingbusiness.org., 2019. [online], available at: https://www.doingbusiness.org/content/dam/doingBusiness/media/Annual-Reports/English/DB2019-report_web-version.pdf.

ethical sense. It can also be a 'responsible for' in a causal mode; some equate it with a charitable contribution or socially conscious. A few see a sort of fiduciary duty imposing higher standards of behaviour on businessmen than on citizens at large".[57] Carroll has held that corporate social responsibility (CSR) covers four categories, namely the economic, legal, ethical and discretionary.[58] The economic aspect views CSR as responsible for job creation, remunerations in the form of payment of wages or salaries and similar commitments. The legal aspect views CSR as business enterprises have to comply with all legal provisions imposed on their activities. The ethical aspect views CSR in the light of companies sticking to ethical standards in their overall actions as a supplement to their legal mandate.[59]

4.6.4 Supplier Development Programmes

The Organization for Economic Cooperation and Development (OECD) adopted guidelines for corporate social responsibility.[60] The guidelines emphasise two mechanisms for supplier development programmes. First is encouraging local capacity building through close cooperation with the local community.[61] This entails business interests, developing the enterprise activities in the local and foreign markets in line with the current commercial practice.[62] The development of human capital formation, creating employment opportunities and facilitating training opportunities for local employees.[63] Furthermore, Suave

[57]Votaw, D., 1972. Genius Becomes Rare: A Comment on the Doctrine of Social Responsibility Pt. I. *California Management Review*, 15(2), pp. 25–31.

[58]Carroll, A.B., 1979. A Three-Dimensional Conceptual Model of Corporate Performance. *Academy of Management Review*, 4(4), pp. 497–505.

[59]Ibid.

[60]Rodrik, Dani. 2008. Normalizing Industrial Policy. Commission on Growth and Development Working Paper; No. 3. World Bank, Washington, DC. © World Bank. https://openknowledge.worldbank.org/handle/10986/28009. License: CC BY 3.0 IGO.

[61]The OECD Guidelines and Other Corporate Responsibility Instruments. https://www.oecd.org/corporate/mne/WP-2001_5.pdf.

[62]Ibid.

[63]The OECD Guidelines and Other Corporate Responsibility Instruments. https://www.oecd.org/corporate/mne/WP-2001_5.pdf.

states that "requirements to implement an approved supplier development programme represent an attractive alternative to enforcing local content rules. Such programmes should: foster adoption of new technologies by the suppliers and produce measurable improvements in value-added, upgrade managerial and technical skills, achieve measurable improvements in managerial practices, achieve measurable improvements in quality output and quality processes and management, result in the insertion of local suppliers into global value chains (GVCs), domestically at least and preferably within export markets over time".[64]

4.6.5 Selected Industrial Policy Measures

These are selective industrial policies that implement particular barriers that affect a given sector or sub-sector of the economy. Rodrick states that the use of selecting policies is usually likened to "picking winners".[65] Moran argues that "a key aspect of needed institutional adaptation involves the recognition that focus of the government should not be on picking winners".[66] Moran concluded that the private sectors and other notable stakeholders could design and evaluate effective strategic collaborations. Suave argued that "because policymakers do not have full knowledge about the activities that need the most support. It is best to design flexible policies with well-specified objectives able to improve overall allocative efficiency, operate within an agreed timeline (including sunset provisions) evolve key performance indicators and clearly defined exit strategies when policies reveal themselves ineffective".[67] The outstanding target industry support is the following:

[64]Sauvé, P., 2016. Life Beyond Local Content: Exploring Alternative Measures of Industry Support in the Context of WTO Accession. *Journal of International Trade*, 1, pp. 1–28.

[65]Rodrik, Dani, 2008. Normalizing Industrial Policy. Commission on Growth and Development Working Paper; No. 3. World Bank, Washington, DC. © World Bank. https://openknowledge.worldbank.org/handle/10986/28009. License: CC BY 3.0 IGO.

[66]Moran, T., 2014. Foreign Investment and Supply Chains in Emerging Markets: Recurring Problems and Demonstrated Solutions. Peterson Institute for International Economics Working Paper (2014-12).

[67]Sauvé, P., 2016. Life Beyond Local Content: Exploring Alternative Measures of Industry Support in the Context of WTO Accession. *Journal of International Trade*, 1, pp. 1–28.

- **Promotion of Labour and Skills**: The development of this can be through apprenticeships, training partnerships with international companies, foreign scholarships to address the skill gaps, continuous collaborative strategies for education and research between different sectors and academic institutions; firm supplier joint developmental programmes, capacity building for enhanced labour market forecasting.
- **Technological advancement**: Development of the emergence of firm clusters, granting R&D subsidies and grants for innovative projects proposed by local companies and carried out by local research institutions; awarding prizes and tax benefits to innovative local firms; facilitating a long-term collaborative strategy for education and research between firms, developing production subsidy to innovative local firms.
- **Investment incentives**: Such incentives can be directed to attracting innovative lead firms in the target sectors. This increases competition in the local market and facilitate local suppliers to increase performance.
- **Systems and institutions**: A country can implement competitive industrial strategies, cluster policies, and strategic planning with industry. Experience from other economies proposes the centrally important role of investment promotion agencies and of special economic zones that have a mass human capital, business services and logistics.

4.7 Local Content Policies in the Oil and Gas Sector

Local content policies were first adopted and implemented in the 1970s in Norway's Continental Shelf.[68] Local content was used to facilitate the bridging of the skills and technology gap between the prominent

[68] Heum, P., 2008. Local Content Development: Experiences from Oil and Gas Activities in Norway. https://openaccess.nhh.no/nhh-xmlui/bitstream/handle/11250/166156/A02_08.pdf?sequence=1&isAllowed=y.

multinational companies and the local oil and gas companies. Local content found its provisions entrenched in the licensing regime, contractual framework, legislation and different policy statements in Norwegian oil and gas economy. The policy was a means on leveraging on the existing skills in Norway, predominantly from the shipbuilding industry into the nascent oil and gas industry. Local content policies were later transplanted to the Nigerian oil and gas industry. Whereby the policy first started from the point of indigenization. A typology of local content policy similar to the community content whereby there is a single unit to benefit from the oil and gas resources.

In the Sub-Saharan local content's development trajectory, these policies have failed to live up to the expectations of fixing market failure, encouraging and facilitating local inclusion, bridging the technological gap, meeting the demands of the oil and gas industry. Local content policies have so far been exemplary in other countries apart from Sub-Saharan countries. This is attributed solely to the fact that oil and gas policies cannot thrive in a vacuum other factors need to be analysed to ascertain the full effect of local content policies in oil and gas economies. It has been argued that local content policies must develop linkages from the oil and gas sector to other sectors for there to be robust economic changes in any given economy.

4.8 Linkages in the Oil and Gas Industry

Hirschman stated that "these industry linkages were essential for economic development and focused on how to promote the formation of robust input markets in poor countries and target investment to the industries with the strongest linkages".[69] He further categorises and defines linkages as "backward links relate to the demand of one industry for the outputs of other industries (to be used as its inputs); forward links relate to the output of one industry as supplied to other industries (as input for their output), and financial links relate to the taxes paid by the industry to the government (which can then be spent

[69] Hirschman, Albert O., 1958. *The Strategy of Economic Development*. Yale University Press.

on other goods and services or saved)"[70] Marcel and Tissot state that "the purpose of developing national content policies is to increase the value generated by the exploitation of a resource that remains in the domestic economy; and to develop linkages between the oil sector and the rest of the economy. What matters is not only what happens in the sector, but what happens beyond".[71] The GIZ mentions the different linkages and these are[72]: Fiscal linkages refer to the payments such as royalties that the extractive industry companies make to governments to extract and sell the resources. Production linkages refer to the goods and services that can be developed as a result of the extractive industry operations. These can be divided into the following: Upstream (or backward) linkages: that relate to the procurement of goods and services that the extractive industry sector requires to operate. For example, this includes welding services needed to maintain mining equipment.

Downstream (or forward) linkages: that relate to the beneficiation of extracted commodities through refining, smelting and further downstream processing of the commodity before reaching the final consumer. For example, 98% of the world's mined iron ore is used to make steel and therefore needs to pass through a steel mill. Horizontal (or lateral) linkages: that relate to the development of new industries using the capabilities of the extractive industry-related supply chain. Apart from originating from the extractive industry project, horizontal linkages can occur from the upstream and downstream level. For example, mining trucks could be re-engineered and adapted to service the logging industry. Consumption linkages: relate to the demand for goods and services resulting from the spending of earnings from the extractive industry sector. This effect is also referred to as induced employment. For example, when a truck driver employed to work at a mine site spends his salary on buying construction materials to build a family house. Side-stream linkages (or enabling factors): relating to the supporting services,

[70]Ibid.

[71]Marcel, V., Tissot, R., Paul, A. and Omonbude, E., 2016. *A Local Content Decision Tree for Emerging Producers*. London, UK: Chatham House, The Royal Institute of International Affairs.

[72]Linkages to the Resource Sector. The Role of Companies, Government and International Development Cooperation. The Giz. http://ccsi.columbia.edu/files/2016/07/Linkages-to-the-resource-sector-GIZ-CCSI-2016.pdf.pdf.

know-how and infrastructure necessary for the extractive industry value chain to function. These can benefit the broader economy and can be sub-divided into Knowledge (technological) linkages: relate to the transfer of knowledge and technological know-how. For example, a mining company may require IT services with higher expertise. Apart from benefitting the mining company itself, the developed IT expertise may also benefit other sectors in the economy. Spatial (infrastructure) linkages: relate to the benefits associated with the infrastructure developed for an extractive industry project profiting other actors in the economy. For example, an iron-ore mine is likely going to require railway infrastructure. Other companies (from the mining sector and non-mining sectors) and passengers will benefit from this railway infrastructure if they are granted access.

The above typologies are all critical in facilitating economic development in any local economy. UNCTAD concludes that lack of linkages and limited absorptive capacity in local industries and host economies has led to limited local development.[73] Linkages do not develop in a vacuum, and the government must devise strategies and capabilities to channel these linkages. The government must develop and adopt some policies that will facilitate the effectiveness of linkages. The African Development Bank holds the view that backward linkages to suppliers and service providers are common, and they have a higher development potential than forwarding linkages to processors and distributors.[74] Mbayi also supports Korineck's view and states that "indeed, most of the evidence of linkage broadening and deepening from African extractives comes from backward linkages although there also are examples of successful forward linkage formation, the prime example being Botswana's diamond-polishing industry".[75] UNCTAD states that "host governments have devised specifics linkage instruments and strategies

[73]UNCTAD. 2013. *Commodities and Development Report: Perennial Problems, New Challenges and Evolving Perspectives*. United Nations Publication.

[74]ADB (African Development Bank), 2013. African Economic Outlook: Structural Transformation and Extractives. African Development Bank, the Development Centre of the Organisation for Economic Co-operation and Development and United Nations.

[75]Mbayi, L., 2011. Linkages in Botswana's Diamond Cutting and Polishing Industry. MMCP Discussion Paper. The Open University and University of Cape Town.

to promote linkages in the extractive sector, e.g. ownership requirements, local content requirements, local processing standards, hiring requirements, mandatory corporate social responsibility programmes, supplier development programmes".[76] All the mentioned linkage intervention mechanisms can be attributed to local content requirements and objectives.

4.9 Factors that Promote and Constrain Linkage Promotion

Hansen states the factor that either promotes or constrains linkage development within an economy: (1) Government strategies and capabilities (2) multinational corporations (MNCs) strategies and capabilities (3) the local industrial strategies and capabilities (4) donor strategies and influence.[77] Hansen further argues that "forming local linkages is part of foreign firms' strategies to reduce costs, increase efficiency, access resources and capabilities, and reduce risks".[78] The promotion and restrain placed on linkages formation and development is mainly due to governmental intervention. The factors elaborated below show what promotes as well as constrain the development of linkages in various jurisdictions.

- **Government Strategies:**

Tordo and Anouti state that "usually companies that supply the oil and gas sector are big foreign firms that have enough market power to displace small local firms from the value chain of the sector".[79] Furthermore, they tend to have well-established and long-term relationships

[76]UNCTAD. 2010. Creating Business Linkages: A Policy Perspective. Geneva, UNCTAD.

[77]Hansen, M.W., 2014. From Enclave to Linkage Economies? A Review of the Literature on Linkages Between Extractive Multinational Corporations and Local Industry in Africa (No. 2014: 02). DIIS Working Paper.

[78]Ibid.

[79]Tordo, S., Warner, M., Manzano, O. and Anouti, Y., 2013. *Local Content Policies in the Oil and Gas Sector*. The World Bank.

with oil and gas companies. This may create a bias against the use of local suppliers similar to the distortions generated by the practice of dumping in international trade. The Brazilian government in 2009 came up with the Site Opportunities Supply Chain of Petroleum and Natural Gas. This initiative facilitated market information about the demand for materials, equipment and components required to carry out oil and gas projects. This shows that local content requirements can address market failure and information asymmetries.

- **Foreign Firms' Strategies and Capabilities**:

Multinational corporations' strategies and capabilities have paved the way for linkage formation in African extractive economies. These capabilities have been argued by Hoekman and Martin stating that "lead extractive foreign firms will be able to dictate supplier firms and governments to their terms".[80] Such as outsourcing mechanisms or value chain disintegration tactics. Hansen and Schaumburg-Muller state that "outsourcing takes place to reduce costs, spread risks, and obtain benefits of specialization, and tap into the resources and capabilities of other firms".[81] Firms can develop shortlists of competent suppliers, they organise competitive bids for contracts, and they devise supplier development and training programmes. Furthermore, they work with authorities to facilitate education, training and local technical capacity. Foreign firms can promote linkages formation and constrain linkage formation

- **Local Industries Strategy and Donor Influence**:

Local industry strategies and capabilities are also a driver for linkage formation. McKinsey states that "the financial, strategic and technological capabilities of African industries have improved significantly in

[80]Hoekman, B., and Martin, W., 2012. Reducing Distortions in International Commodity Markets: An Agenda for Multilateral Cooperation. World Bank Policy Research Working Paper Series.

[81]Hansen, M. W., and Schaumburg-Müller, H. (Eds.), 2006. *Transnational Corporations and Local Firms in Developing Countries Linking and Upgrading*. CBS Press.

recent years".[82] However, problems are in the technical abilities and raising capital are among the significant setbacks of linkage formation in African economies. Based on an expert survey. The African Development Bank shows that technological complexity and lack of skills in local industries places a considerable barrier to linkage development in African extractive economies.[83]

- **Donor Influence:**

Donors have been increasing in the African continent to engage in the extractive sector due to the substantial commercial aspect of the extractive foreign direct investment. Donors can be active in the following activities (1) Provide technical assistance across developing countries; (2) Build infrastructural, institutional and absorptive capacity; (3) Facilitate specific oil and gas investment projects; (4) Intervening between international extractive firms and local governments. Through these initiatives, donor participation can facilitate or constrain linkage development in extractive economies.[84]

4.10 New Trends of Local Content Policies

Local content policies compliance has gained popularity in oil and gas countries. As mentioned earlier, the policies were first introduced in the Norwegian continental shelf in the 1970s. The shift that has been brought about by local content policies is the shift from ad hoc social corporate responsibility towards a prescribed form of implementing local content policies that are strict in definition and qualitative requirements towards value-based economic development.

[82]McKinsey Global Institute, 2011. *Lions on the Move: The Progress and Potential of African Economies*. McKinsey.

[83]ADB (African Development Bank), 2017. African Economic Outlook: Structural Transformation and Extractives. African Development Bank, the Development Centre of the Organisation for Economic Co-operation and Development and United Nations.

[84]Morris, M., Kaplinsky, R. and Kaplan, D. 2011. One Thing Leads to Another Commodities, Linkages and Industrial Development—A Conceptual Overview. MMCP Discussion Paper. Milton Keynes and Cape Town, The Open University and University of Cape Town.

The driving factors to this shift in the adoption and implementation of production development polices, is driven by four main reasons. The scarcity of oil and gas resources, energy transition from fossil fuels to cleaner energy, the effects of COVID-19 pandemic on value chains and increased need for governmental control and intervention in the oil and gas industry to crack through the enclave. The process adopted to meet all these concerns is called government intervention.[85] But with the supervening realities that the resource-rich governments and their local companies cannot meet the demands of the sophisticated, capital-intensive and technological demanding energy sector. This has led to joint ventures between the foreign and local entities through local content policies. Such bottlenecks remove the legitimacy of oil and gas operations within any given community, and this brings about the NYMBISM sentiment that hinders the development of these resources. Thus government intervention measures must be associated with energy justice to legitimatise the extraction and development of oil and gas resources within any society.

But it has been evident in countries like Nigeria, Angola, Chad, Venezuela, Ghana and Brazil that he sole implementation of local content policies does not materialise the local content rhetoric.[86] Intervention is also due to the domestic and international pressure that governments give into increase local participation through strong independent institutions.[87] Legitimatising local content policies are complicated since policymakers only implement these policies due to an expected outcome. Hence the qualitative targets are mostly not met, and the policies are usually transformed into a vehicle of driving corruption and distortion of the local market forces. Norway avoided being dependent on oil rents by creating healthy boundaries between the oil wealth

---

[85] Tordo, S., Warner, M., Manzano, O. and Anouti, Y., 2013. *Local Content Policies in the Oil and Gas Sector*. The World Bank.

[86] Ovadia, J.S., 2016. Local Content Policies and Petro-Development in Sub-Saharan Africa: A Comparative Analysis. *Resources Policy*, 49, pp. 20–30.

[87] Gustafsson, M.T. and Scurrah, M., 2019. Unpacking the Extractivist State: The Role of Weak State Agencies in Promoting Institutional Change in Peru. *The Extractive Industries and Society*, 6(1), pp. 206–214.

and the political systems. At the same time, avoiding market distortions and adopted viable depletion policies that allowed the industry to develop gradually.[88] This was also due to relatively stable socio-economic and political climate in Norway. Unlike other oil and gas resource-rich countries, Norway had a robust fishing and shipping industry that the government did not need to rely on the oil and gas revenues to alleviate poverty in any particular political regime. The UK did not adopt the Norwegian model of avoiding government intervention into the oil and gas industry.[89] The UK employed US oil and gas service firms to run the oil and gas industry. As the government gradually taxed these US companies as well as develop their local capabilities to integrate the British oil and gas industry. The UK also set up the "Offshore Supplies Office" intending to raise British participation from 30 to 70%.[90]

Both the UK and Norway adopted local content policies in the early 1970s after the discovery of oil and gas resources in the North Continental Shelf (NCS). But later in the 1990s due to the Norwegian and UK membership in the European Union local content provisions had to be dropped since EU policies prohibited preference of local firms.[91] As a result, both Norway and the UK created standardised oil and gas practices that allowed the local firms and workforce to compete within the oil and gas industry without having preferential provisions. Brazil local content was among the most stringent local content requirements found in most oil and gas countries. This has led to Brazil's local content policy be a subject of much debate. The policy was introduced in the fifth licensing round in 2003, the policy had heavy requirements, along with hefty fines for non-compliance that increased over other subsequent

[88]Thurber, M.C., Hults, D.R. and Heller, P.R., 2011. Exporting the "Norwegian Model". The Effect of Administrative Design on Oil Sector Performance. *Energy Policy*, 39(9), pp. 5366–5378.

[89]Mendes, P.A., Hall, J., Matos, S. and Silvestre, B., 2014. Reforming Brazil's Offshore Oil and Gas Safety Regulatory Framework: Lessons from Norway, the United Kingdom and the United States. *Energy Policy*, 74, pp. 443–453.

[90]Ibid.

[91]Hufbauer, G.C., Schott, J.J. and Cimino-Isaacs, C., 2013. *Local Content Requirements: A Global Problem* (Vol. 102). Columbia University Press.

rounds.[92] This led to delays in developing Brazil's offshore resources.[93] It resulted in restraining the domestic industry that the policy was meant to foster in the first place. But later the policy became more progressive and steered further investment in Brazil's oil and gas sector by making changes in both the local content policy and the fiscal regime from a tricky policy to a more progressive investor-friendly policy.[94]

In 2017, the Brazilian National Petroleum, Natural Gas and Biofuels Agency (ANP) redrafted the terms of the Brazilian local content policy. In 2018, the Brazilian National Council for Energy Policy (CNPE) and the ANP published a resolution that paved the way for companies to substitute their local content commitments provided in their existing contracts for lower and simpler local content requirements attached with reduced fines.[95] The Norwegian model is still ideal as it advocates for gradual government intervention without distorting the market forces. This allows the oil and gas industry of any jurisdiction to progressively develop into a powerhouse that can propel and stir any economy into economic development. Most countries do not have the features that Norway had. While finding its oil and gas resources, it was a stable democracy, with a predictable economic climate, political stability and uniformity in ideology towards the oil and gas industry even though there was a change in government. But what can be adopted from the Norwegian model, especially in the new trends of climate change and low oil prices countries need to adopt a more flexible value-based approach rather than strict qualitative requirements as Brazil did. The formation of local content policies must recognise and take into account all stakeholders involved to meet the principle of justice as recognition. The

[92]de Almeida, E. and Martinez-Prieto, D., 2015. The Impact and Effectiveness of Local Content Policy on Oil Exploration and Production in Brazil. In Oxford Energy Forum (Vol. 98, No. November 2014).

[93]de Almeida, E. and Martinez-Prieto, D., 2015. The Impact and Effectiveness of Local Content Policy on Oil Exploration and Production in Brazil. In Oxford Energy Forum (Vol. 98, No. November 2014).

[94]Araujo, F.C. and Leoneti, A.B., 2019. How Attractive Is Brazil's Oil and Gas Regulatory Framework to Investors? The Extractive Industries and Society.

[95]de Almeida, E. and Martinez-Prieto, D., 2015. The Impact and Effectiveness of Local Content Policy on Oil Exploration and Production in Brazil. In Oxford Energy Forum (Vol. 98, No. November 2014).

implementation and adoption of local content policies must follow due process in the energy value chain to meet the principle of procedural justice. Lastly, the adoption and implementation of local content policies must distribute the gains and ills of having oil and gas within their vicinity.

4.11 Energy Justice Gap Analysis

Energy justice in this book is being intertwined with local integration in the oil and gas industry. Under energy justice, local content policies should not be measured merely on meeting the qualitative requirements within a given sector. But the value of the effectiveness of energy justice must be delivered in the local content application.[96] In adopting and implementing local content policies, several aspects must be taken into accounts such as pace of resource exploitation, institutional responsibilities, degree of regulatory intervention and the coherence with other policies such as industrial policy market regulation educational policies, investment mechanisms, labour, taxation and trade, etc.[97] The aspects mentioned are expected to be efficient and effective in their implementation due to factors such as a country having good governance, industrial capabilities, technological advancement, favourable business environment, as well as favourable political climate. Oil and gas countries have predominantly adopted local content policies. But these policies have not been efficient in their application. Norway has been the only country with good local content since its inception in the mid-1970s. The research on Norway does prove that the country had a favourable position before finding the oil and gas resources. Good governance of the oil and gas resources was the primary factor for the success of the Norwegian energy sector. But the sovereign funds have also been instrumental in availing economic development in Norway. The Norwegian sovereign wealth fund has over US$1 trillion in assets, and this includes 1.5%

[96]Heffron, R.J., McCauley, D. and Sovacool, B.K., 2015. Resolving Society's Energy Trilemma Through the Energy Justice Metric. *Energy Policy*, 87, pp. 168–176.

[97]Tordo, S., Warner, M., Manzano, O. and Anouti, Y., 2013. *Local Content Policies in the Oil and Gas Sector*. The World Bank.

of global stocks and shares.[98] This makes it the most massive wealth fund in the world. These economic resources have been capitalised to develop Norway. In the UK, the oil and gas industry is worth over £40 Billion and employs the local workforce into the industry, with 45% of the oil and gas jobs being obtained in Scotland.[99] Though the UK at first employed US oil servicing companies but later developed a robust oil and gas industry with strong local players such as British Petroleum. The above examples of Norway and the UK show that local content has been solely based on economic development. As long as a country benefits from its natural resource endowments, local content policies seem to be commendable in this regard. As part of the energy justice gap, the research identifies that there is limited incorporation of justice in local content policies. Furthermore, though other countries do not have the same circumstances as Norway, Brazil and the UK if energy justice is adequately applicable, this can lead to the same results due to the following reasons.

4.11.1 Distributional Justice

Energy justice is spatial, temporal as well as a societal concept that involves the unequal allocation of benefits and ills and the uneven distribution of the associated responsibilities.[100] The energy value chain is also disintegrated in the sense that the remote areas where the oil and gas are found are not the areas where the energy resource business is undertaken. Investments are not poured in rural energy-rich regions but in cities with proper infrastructure and capability to absorb the foreign investments. In these rural energy resource areas, only a lack worth of investments that gives a social license to operate and facilitates the energy industry is undertaken. Hence local content provisions cannot be efficient in this manner since there is an uneven distribution of ills and benefits.

---

[98] Bahoo, S., Alon, I. and Paltrinieri, A., 2019. Sovereign Wealth Funds: Past, Present and Future. *International Review of Financial Analysis*, 67, p. 101418.

[99] Littlechild, S. and Vaidya, K., 2019. *Energy Strategies for the UK*. Routledge.

[100] Walker, G., 2009. Beyond Distribution and Proximity: Exploring the Multiple Spatialities of Environmental Justice. *Antipode*, 41(4), pp. 614–636.

4.11.2 Procedural Justice

Energy justice calls for the use of an exemplary process that equally engages all stakeholders without discriminating any stakeholders involved.[101] This principle calls for all stakeholders to have access to adequate information in a timeous manner. Local content policies have been used to support the interests of the elite. The energy industry is closely linked to politics due to its lucrative advantage compared to other sectors of the economy. Local content policies create rent-seeking tendencies by those connected to the elite, and local content policies have deliberate preferences towards specific stakeholders (mostly local stakeholders). This does not meet the principle of procedural justice; thus, local content policies cannot thrive without due diligence.

4.11.3 Justice as Recognition

Justice, as recognition, takes into account the cultural, social and political aspects of a community.[102] Recognition is not to be put in the same place as participation. Local content calls for the local participation of the local population into the oil and gas value chain. But in this instance, local content will have to take into account the particular circumstances of a given community (most probably the energy resource-rich community). Developing national content may not encompass the peculiarities of energy resource-rich communities. The paradox reached is that a community content serves the principles of justice as recognition. But a national content does not, since the oil and gas resources are mostly found in isolated remote communities.

[101] Bullard, R., 2000. Environmental Justice in the 21st Century. People of Colour Environmental Groups. Directory, pp. 1–21.

[102] Walker, G., 2009. Beyond Distribution and Proximity: Exploring the Multiple Spatialities of Environmental Justice. *Antipode*, 41(4), pp. 614–636.

4.12 Summary

This chapter has provided a clear literature review on local content policies both as an abstract notion and in its real-life implementation. The different competing definitions of local content policies across different economies. Explained within different theoretical constructs, namely energy justice, localism and regionalism. It is from these constructs that local content policies typologies are found, namely the national content policy, the community content policy and the regional content policy. The alternatives of local content policies show that these policies are hefty on the balance sheet of the investors and the trade-offs are also severe on the local community. That is why we must address the constraints and promoting factors of linkage development. Because only through linkage development may Sub-Saharan countries be able to crack through the enclave of the oil and gas sector. The enclave nature of this sector is the primary cause of Sub-Saharan countries having resource curse and minimal industrial development from the oil and gas industry. Compared to developed countries such as Brazil and Norway.

References

A Guide to Getting Started in Local Procurement: For Companies Seeking the Benefits of Linkages with Local SMEs. International Financial Corporation. http://www.ifc.org/wps/wcm/connect/03e40880488553ccb09cf26a6515bb18/IFC_LPPGuide_PDF20110708.pdf?MOD=AJPERES.

ADB (African Development Bank), 2013. African Economic Outlook: Structural Transformation and Extractives. African Development Bank, the Development Centre of the Organisation for Economic Co-operation and Development and United Nations.

Agreement on the Exploration, Development and Production Sharing for the Shah Deniz Prospective Area in the Azerbaijan Sector of the Caspian Sea, 1996, http://www.bp.com/liveassets/bp_internet/bp_caspian/bp_caspian_en/STAGING/local_assets/downloads_pdfs/pq/SD_PSA.pdf.

Araujo, F.C. and Leonetti, A.B., 2019. How Attractive Is Brazil's Oil and Gas Regulatory Framework for Investors? The Extractive Industries and Society.

Bahoo, S., Alon, I. and Paltrinieri, A., 2019. Sovereign Wealth Funds: Past, Present and Future. *International Review of Financial Analysis*, 67, p. 101418.

Briffault, R., 2000. Localism and Regionalism. *Buffalo Law Review*, 48, p. 1.

Bullard, R., 2000. Environmental Justice in the 21st Century. People of Colour Environmental Groups. Directory, pp. 1–21.

Carroll, A.B., 1979. A Three-Dimensional Conceptual Model of Corporate Performance. *Academy of Management Review*, 4(4), pp. 497–505.

Cimino, Cathleen, Hufbauer, Gary Clyde and Schott, Jeffrey J., A Proposed Code to Discipline Local Content Requirements. Peterson Institute for International Economics Policy Brief No PB14–6, February 2014, p. 1.

Corden, W.M., 1985. *Protection, Growth and Trade: Essays in International Economics*. Oxford: Blackwell.

De Almeida, E. and Martinez-Prieto, D., 2015. The Impact and Effectiveness of the Local Content Policy on Oil Exploration and Production in Brazil. In Oxford Energy Forum (Vol. 98, No. November 2014).

Doingbusiness.org., 2019. [online] Available at: https://www.doingbusiness.org/content/dam/doingBusiness/media/Annual-Reports/English/DB2019-report_web-version.pdf.

Employment of Foreign Citizens Non-Resident, Decree 5/95, Ministry of the Interior, Republic of Angola. http://www.sme.ao/attachments/article/209/Decree%20No.%20595%20of%20April%207%20%20Employment%20of%20Foreign%20Citizens%20Non-resident.pdf.

Grant, R.M., 1991. Porter's 'Competitive Advantage of Nations': An Assessment. *Strategic Management Journal*, 12(7), pp. 535–548.

Grossman, G.M., 1981. The Theory of Domestic Content Protection and Content Preference. *The Quarterly Journal of Economics*, 96(4), pp. 583–603.

Gustafsson, M.T. and Scurrah, M., 2019. Unpacking the Extractivist State: The Role of Weak State Agencies in Promoting Institutional Change in Peru. *The Extractive Industries and Society*, 6(1), pp. 206–214.

Haas, Ernst B., 1964. *Beyond the Nation-State*. Stanford: Stanford University Press.

Hansen, M.W., 2014. From Enclave to Linkage Economies? A Review of the Literature on Linkages Between Extractive Multinational Corporations and Local Industry in Africa (No. 2014: 02). DIIS Working Paper.

Hansen, M.W. and Schaumburg-Müller, H. (Eds.), 2006. *Transnational Corporations and Local Firms in Developing Countries Linking and Upgrading*. CBS Press.

Heffron, R.J. and McCauley, D., 2017. The Concept of Energy Justice Across the Disciplines. *Energy Policy*, 105, pp. 658–667.

Heffron, R.J., McCauley, D. and Sovacool, B.K., 2015. Resolving Society's Energy Trilemma Through the Energy Justice Metric. *Energy Policy*, 87, pp. 168–176.

Hettne, B. and Söderbaum, F., 1998. The New Regionalism Approach. *Politeia*, 17(3), pp. 6–21.

Heum, P., 2008. Local Content Development: Experiences from Oil and Gas Activities in Norway. https://openaccess.nhh.no/nhhxmlui/bitstream/handle/11250/166156/A02_08.pdf?sequence=1&isAllowed=y.

Hirschman, Albert O., 1958. *The Strategy of Economic Development*. Yale University Press.

Hoekman, B. and Martin, W., 2012 Reducing Distortions in International Commodity Markets: An Agenda for Multilateral Cooperation. World Bank Policy Research Working Paper Series.

Hufbauer, G.C., Schott, J.J. and Cimino-Isaacs, C., 2013. *Local Content Requirements: A Global Problem* (Vol. 102). Columbia University Press.

Kessler, Anke S., Hansen, Nico A. and Lessmann, Christian., 2011. Interregional Redistribution and Mobility in Federations: A Positive Approach. *Review of Economic Studies*, 78, p. 1345.

Kolstad, I. and Kinyondo, A., 2017. Alternatives to Local Content Requirements in Resource-Rich Countries. *Oxford Development Studies*, 45(4), pp. 409–423.

Linkages to the Resource Sector. The Role of Companies, Government and International Development Cooperation. The Giz. http://ccsi.columbia.edu/files/2016/07/Linkages-to-the-resource-sector-GIZ-CCSI-2016.pdf.pdf.

Littlechild, S. and Vaidya, K., 2019. *Energy Strategies for the UK*. Routledge.

Lloyd, P.J., 1973. *Non-tariff Distortions of Australian Trade*. Canberra: Australian National University Press.

Marcel, V., Tissot, R., Paul, A. and Omonbude, E., 2016. *A Local Content Decision Tree for Emerging Producers*. London, UK: Chatham House, The Royal Institute of International Affairs.

Mbayi, L., 2011 Linkages in Botswana's Diamond Cutting and Polishing Industry. MMCP Discussion Paper. The Open University and University of Cape Town.

McKinsey Global Institute, 2011. *Lions on the Move: The Progress and Potential of African Economies*. McKinsey.

Mendes, P.A., Hall, J., Matos, S. and Silvestre, B., 2014. Reforming Brazil's Offshore Oil and Gas Safety Regulatory Framework: Lessons from Norway, the United Kingdom and the United States. *Energy Policy*, 74, pp. 443–453.
Moran, T., 2014. Foreign Investment and Supply Chains in Emerging Markets: Recurring Problems and Demonstrated Solutions. Peterson Institute for International Economics Working Paper (2014-12).
Morris, M., Kaplinsky, R. and Kaplan, D., 2011. One Thing Leads to Another Commodities, Linkages and Industrial Development—A Conceptual Overview. MMCP Discussion Paper. Milton Keynes and Cape Town, The Open University and University of Cape Town.
Nwapi, C., 2015. Defining the "Local" in Local Content Requirements in the Oil and Gas and Mining Sectors in Developing Countries. *Law and Development Review*, 8(1), pp. 187–216.
Ovadia, J.S., 2012. The Dual Nature of Local Content in Angola's Oil and Gas Industry: Development vs Elite Accumulation. *Journal of Contemporary African Studies*, 30(3), pp. 395–417.
Ovadia, J.S., 2016. Local Content Policies and Petro-Development in Sub-Saharan Africa: A Comparative Analysis. *Resources Policy*, 49, pp. 20–30.
Petroleum (Local Content and Local Participation) Regulations, 2013. LI 2204, Reg 49. Available at http://www.reportingoilandgas.org/wp-content/uploads/PETROLEUMLOCAL-CONTENT-AND-LOCAL-PARTICIPATION-REGULATIONS2013.pdf.
Presidential Regulation No. 54/2010 and The Oil and Gas Law No.22/2001 Segregating Regulatory Roles and Commercial Operations.
Rodrik, Dani, 2008. Normalizing Industrial Policy. Commission on Growth and Development Working Paper; No. 3. World Bank, Washington, DC. © World Bank. https://openknowledge.worldbank.org/handle/10986/28009. License: CC BY 3.0 IGO.
Sauvé, P., 2016. Life Beyond Local Content: Exploring Alternative Measures of Industry Support in the Context of WTO Accession. *Journal of International Trade*, 1, pp. 1–28.
The OECD Guidelines and Other Corporate Responsibility Instruments. https://www.oecd.org/corporate/mne/WP-2001_5.pdf.
The United Republic of Tanzania, Local Content Policy (2014).
Thurber, M.C., Hults, D.R. and Heller, P.R., 2011. Exporting the "Norwegian Model": The Effect of Administrative Design on Oil Sector Performance. *Energy Policy*, 39(9), pp. 5366–5378.
Tordo, S., Warner, M., Manzano, O. and Anouti, Y., 2013. *Local Content Policies in the Oil and Gas Sector*. The World Bank.

UNCTAD (2010) Creating Business Linkages: A Policy Perspective. Geneva, UNCTAD.
Votaw, D., 1972. Genius Becomes Rare: A Comment on the Doctrine of Social Responsibility Pt. I. *California Management Review*, 15(2), pp. 25–31.
Walker, G., 2009. Beyond Distribution and Proximity: Exploring the Multiple Spatialities of Environmental Justice. *Antipode*, 41(4), pp. 614–636.
Warner, M., 2007. Community Content: The Interface of Community Investment Programmes with Local Content Practices in the Oil and Gas Development Sector. Overseas Development Institute (ODI) Briefing Note (9).
Warner, M., 2010. Unpacking Local Content Metrics and Measurement Dr Michael Warner, November PDF. [online] Docplayer.net. Available at: https://docplayer.net/48696189-Unpacking-local-content-metrics-and-measurement-dr-michael-warner-november-2010.html.
Warner, M., 2011. *Local Content in Procurement: Creating Local Jobs and Competitive Domestic Industries in Supply Chains*. Sheffield, UK: Greenleaf Publishing.
Warner, M., 2017. *Local Content in Procurement: Creating Local Jobs and Competitive Domestic Industries in Supply Chains*. Routledge.
Wellisz, S. and Findlay, R., 1988. The State and the Invisible Hand. *The World Bank Research Observer*, 3(1), pp. 59–80.

5

Country Case Studies

5.1 Introduction

Local content addresses oil and gas industry bottlenecks. The policy has entrenched into the legislative, regulatory and contractual frameworks of different energy resource-rich countries. Countries are mostly assessed on the policymaker's probable intentions, policy structure and what cause of action was taken before the discovery of oil and gas discovery and after thereof. The case studies have different social, political and economic structures that cannot be put to comparison at the current state of affairs in each country. We undertake a cross-country analysis made across several countries, namely Tanzania, Kenya, Uganda, Nigeria, Brazil, Angola, Mozambique and Norway. This will facilitate the in-depth understanding of whether local content policies are well designed and implemented to crack through the enclave of the oil and gas industry in Sub-Saharan Africa. We will see that in developed economies such as Brazil and Norway, it is easier for these policies to crack through the enclave nature of the oil and gas industry because they are dependent on the oil and gas industry for economic development. It is easier for countries like Norway and Brazil, to have fewer regulations and requirements when it comes to local content policies, unlike countries from

R. S. Muhongo, *Energy Justice*, Energy, Climate and the Environment,
https://doi.org/10.1007/978-3-030-61338-9_5

Sub-Saharan Africa. Unlike Sub-Saharan economies that are dependent on natural resources that rely on "natural-resource based development" unlike developed economies that are characterised by strong manufacturing, technology, service and shipping sector. Something that is not readily available in other developed oil and gas economies. This chapter illustrates what local content policies are in a real-life context according to the countries that have implemented them.

5.2 Country Selection Justification

Norway was selected due to its exemplary and commendable oil and gas industry. Norway has a sector that avoided the resource curse, Dutch disease, and the industry is transparent with efficient regulatory systems. Furthermore, the industry has also developed major local players in the oil and gas industry. Norway has been a poster boy for the development and adoption of local content policies since the proliferation of these policies in oil and gas plus oil and gas-rich jurisdictions. The country offers great lessons to be drawn by nascent oil and gas-rich economies to transform these natural resources into a vehicle leading to economic prosperity for any oil and gas-rich country. As much as Norway is being used in this book for comparable analytical reasons as to why local content policies thrive in Norway and not in other developing countries such as Nigeria and Angola. Norway is also used to determine whether national content policies only thrive in jurisdictions with a stable economic and political background, or whether the transplantation of the Norwegian model can be shipped into other jurisdictions and yield similar results as it did in Norway. Norway had adequate depletion policies, lenient fiscal measures towards the oil and gas investors, concentrated on developing local capabilities and advocated for robust R&D mechanisms are these the most critical aspects for a nascent oil and gas economy like Tanzania, Kenya, Uganda and Mozambique. Norway also provides the ability to analyse between demanding strict qualitative local content requirements and value-based local content requirements. As Norway adopted a more value-based approach, unlike Brazil, that chose strict local content requirements. Such an analysis will facilitate the understanding of what local content policies are to be adopted according to the given prevailing circumstances of a given country. Norway provides the basis of creating both a theoretical analysis based on energy justice into policy as well as a commendable policy analysis for oil and gas countries.

Brazil was selected due to its complicated legal and regulatory regime. The Brazilian industry has developed local content policies that are strict with rigid qualitative demands. The Brazil case study brings about depth understanding into distinctions between value-based and qualitative based local content policy requirements. However, Brazil implemented strict qualitative targets coupled with hefty fines for non-compliance of meeting the local content commitments as agreed between the Brazilian government and the oil and gas multinational corporations. Brazil shows the effects of local content policies on attracting foreign investors, significantly when the policies as strict and later lenient as Brazil later relaxed their local content expectations to attract further oil and gas foreign investment.

Furthermore, the effects of politics on local content policies is also analysed in the Brazilian case study. The view of whether local content policies should adopt a resource nationalistic perspective or a free market perspective in local content policy implementation is also shown in the Brazilian case study. Brazil leads the transition from demanding a strict approach to a value-based approach. And how that affects local oil and gas market participants as well as the role of Petrobras as a monopoly in the oil and gas industry. Brazil also shows the role of corruption in government institutions and how local content policies are used by the political elite. Brazil also indicates the development of local content policies in countries that already have an industrial base before the discovery of oil and gas discoveries. Such an analysis is drawn by showing that a country like Brazil and even Norway could leverage on their shipping and fishing industries that already have global recognition to build local competence for the offshore oil and gas resources.

Nigeria and Angola were selected because these are the first Sub-Saharan African countries to adopt and implement local content policies. Their case study will give us an insight into the development of local content policies in developing countries with low technological advancements, low-quality skilled labour, and minimal linkages between the oil and gas industry and the rest of the economy. Nigeria and Angola also offer a unique insight on understanding the role of local content policies in building a local capitalist elite through corrupt mean but within the legal and regulatory framework of a country. Norway and Brazil are commendable local content countries, but they are far off as countries that a nascent oil and gas economy should draw lessons. Thus, Angola and Nigeria are used in this book to bridge the gap between the developing economies of Brazil and Norway and the nascent oil and gas countries such as Mozambique, Kenya, Tanzania and Uganda. Unlike Brazil and Norway, Nigeria and Angola introduced policies such as the indigenisation policy that gave preferential ownership to the local people after the discovery of oil and gas resources.

Such policies facilitate further research into understanding local content policies that advocate for local ownership of oil and gas resources, especially in developing countries like Kenya, Tanzania, Nigeria and Uganda. A country like Kenya does not have the same leverage on the size of its resources as Uganda, Nigeria and Mozambique. This brings about an understanding of what typology of local content policy could be more efficient and robust in its implementation due to the size of the found resources. Lastly, Nigeria and Angola are selected because their policies show the effects of policy transplantation from developed mature oil and gas producers on developing nascent economies. Sub-Saharan Africa countries lack specialised skills for the oil and gas industry. This poses a massive obstacle for the development of the oil and gas industry in these countries. The study in this book offers a unique opportunity both for policymakers and researchers to develop novel ideas as what type of local content policy is readily available to meet the industry's bottlenecks given the prevailing circumstances. These countries do not have the same circumstances as Norway and Brazil of having robust industrial capacity before the discovery of oil and gas resources. As it may be evident that national content policies do thrive in developed countries, as will be shown in the cross-country analysis. But the presence of Kenya, Tanzania, Mozambique and Uganda is where we analyse if local content policies can efficiently crack through the enclave of the oil and gas industry if adopted using energy justice framework. The fact that Kenya, Uganda and Tanzania all belong to the East African Community, and they have consolidated as well as pooled resources to meet different demands in the past. They offer an understanding of whether regional integration can be used to formulate a regional content policy to leverage local skills, capital and labour against the oil and gas multinational companies. Furthermore, Mozambique, Kenya, Uganda and Tanzania bring about an opportunity of creating a local content policy that is not purely based on economic goals. But also on justice, and this is where the application of energy justice comes into play since these are nascent oil and gas countries.

5.3 The Economy Before Oil and Gas Discovery: A Cross-Country Analysis

The economy before the discovery of oil and gas resources is of paramount importance in understanding why some local content policies thrive in other economies. While when transplanted to other economies regardless of the similarities the results are the total adversary of the latter result. Early in the 1880s, the Norwegian economy was mainly driven by the robust shipping economy and accompanied by massive exports of un-processed dried or salted fish and timber.[1] The fishing sector was characterised by whaling and canning.[2] Later in 1870, Norway's manufacturing sector had steam-driven equipment, sawing equipment etc. Despite these developments, Norway still was regarded as a developing country according to numerous indicators[3].

On the other hand, Brazil also discovered oil and gas resources in the 1800s, the country's economy was also mainly driven by the shipping and fishing industry at the time of discovering the oil and gas

[1]Cappelen, Å. and Mjøset, L., 2009. Can Norway Be a Role Model for Natural Resource Abundant Countries? Chap. 3, pp. 44–72.

[2]Engen, O.A., 2009. The Development of the Norwegian Petroleum Innovation System: A Historical Overview. *Innovation, Path Dependency and Policy: The Norwegian Case*, pp. 179–207.

[3]Andersen, S.S., 1993. *The Struggle Over North Sea Oil and Gas: Government Strategies in Denmark, Britain, and Norway*. A Scandinavian University Press Publication.

resources.[4] Both Norway and Brazil paced their way into the oil and gas industry. This was done by first developing the local capacity and governmental institutions that could integrate the oil and gas value chain.[5] Nigeria discovered its oil and gas resources in the 1960s. Sub-Saharan African countries never had an industrial base, such as the shipping industry. The region's economy was mainly driven by subsistence farming before the discovery of oil and gas resources[6]. Countries like Tanzania, Kenya, Mozambique, Angola, Liberia, Ghana and Uganda are not old oil and gas resource countries like Norway and Brazil. The Sub-Saharan African economies were predominantly characterised by either civil unrest or small scale subsistence farming surviving on rural livelihoods "working in agriculture, livestock, forestry and fisheries, applying traditional manual and low-productivity technology".[7] Other countries like Angola, despite having a relatively diversified economy with vital agricultural and manufacturing sectors before oil and gas discoveries. Shortly after independence in 1975, the country spiralled into a civil war that led to massive displacement of people and destruction of the country's infrastructure and human capital. Unlike Norway and Brazil, Sub-Saharan countries did not have the business environment due to civil war and unrest; furthermore, these countries did not possess an industrial base, a manufacturing sector or infrastructure before the discovery of the oil and gas resources.[8].

---

[4]Overview of the Brazilian Oil and Gas Industry. http://www.offshorecenter.dk/filer/files/Project/Internationalisering/OCD%20report%20Brazil.pdf

[5]Local Content: Goals and Brief History. Rio de Janeiro: ANP. http://www.anp.gov.br/brasil-rounds/round10/ingles/conteudo_local_Obj_e.asp.

[6]Paul, I.A., 2015. A Historical Perspective of Petroleum on Nigeria's Economy Crisis since Independence. *Global Journal of Human-Social Science Research*.

[7]Cruz, A.S. and Mafambissa, F.J., 2016. Industries Without Smokestacks. Industries Without Smokestacks, p. 232.

[8]Maxon, R., 2009. *East Africa: An Introductory History*. West Virginia University Press.

5.4 Initial Government Intervention Mechanisms in the Oil and Gas Industry: A Cross-Country Analysis

As mentioned earlier to have local content policies, there must be a governmental intervention into the oil and gas industry. The need for government intervention is mainly due to five given reasons, such as:

I. The multinational corporation's practice "bundling", a process of putting smaller contracts into a big pile. This practice excludes local small and medium enterprises (SMEs). These SMEs also struggle in the cash traps epidemic whereby they are undercapitalised and cannot meet the high-interest rates that could facilitate their integration in the oil and gas industry.[9]

II. Tenders and opportunities to deal with the multinational corporations are usually made "too late". This creates information asymmetry whereby the local companies lack ample time to plan and prepare for different opportunities.[10]

III. The pre-qualification criteria used by multinational corporations is fundamentally based on international standards. Though these standards facilitate efficiency and sound oil and gas operations. These standards hamper local participation.[11]

IV. The nature of the oil and gas industry emphasises engineering, procurement, and construction contracts. This nature of the oil and gas industry coupled with the pre-qualification nature of the oil and gas industry. Makes it difficult to have local participation.[12]

---

[9]Calignano, G. and Vaaland, T.I., 2018. Local Content in Tanzania: Are Local Suppliers Motivated to Improve? *The Extractive Industries and Society*, 5(1), pp. 104–113.

[10]Ibid.

[11]Calignano, G. and Vaaland, T.I., 2018. Local Content in Tanzania: Are Local Suppliers Motivated to Improve? *The Extractive Industries and Society*, 5(1), pp. 104–113.

[12]Calignano, G. and Vaaland, T.I., 2018. Local Content in Tanzania: Are Local Suppliers Motivated to Improve? *The Extractive Industries and Society*, 5(1), pp. 104–113.

Hence government intervention bridges the gap between the strong international and multinational capabilities of the oil and gas companies against the local market participants. Government intervention is the only means that the local participant and local firm can have good and efficient stakeholder relations with the big multinationals. In Brazil, the President at the time Getulio Vargas passed several decrees that were aimed at nationalising the Brazilian oil and gas sector.[13] The state was in control and obtained a monopoly in the market. In Norway, the government did not straight away intervene in the oil and gas industry. The Norwegians paced their way into the industry by first developing skills through joint-venture operations between the local Norwegian companies and the US oil and gas companies.[14] In Nigeria, through the Nigerian Enterprise Promotion Decree (1972); the government adopted and implemented what was known as the indigenisation policy.[15] The policy transferred the ownership of the oil and gas resources from the multinationals to the local oil and gas market participants. Nigeria was not alone in implementing "indigenisation policies". Angola passed Decree 20/82 on the Angolanization of the oil and gas industry—a law that targeted integrating large numbers of Angolans into the oil and gas industry, cracking through the enclaved oil and gas industry.[16] In Kenya, Mozambique, Tanzania and Uganda, the oil and gas industry is nascent. With the economic backbone of these countries being the agro-sector. It is undoubtedly a fact that government intervention is needed to devise a mechanism to integrate the oil and gas industry. The East African countries adopted and implemented local content policies through Exploration Development and Production Act in their countries, respectively, as the provisions for local content are divided between ownership and local sourcing of goods and services.

[13]Overview of the Brazilian Oil and Gas Industry. http://www.offshorecenter.dk/filer/files/Project/Internationalisering/OCD%20report%20Brazil.pdf.

[14]Heum, P., 2008. Local Content Development: Experiences from Oil and Gas Activities in Norway.

[15]Mohammed, I., 1985. The Nigerian Enterprises Promotion Decrees (1972 and 1977) and Indigenisation in Nigeria (Doctoral dissertation, University of Warwick).

[16]Tordo, S., Warner, M., Manzano, O. and Anouti, Y., 2013. *Local Content Policies in the Oil and Gas Sector*. The World Bank.

5.5 Local Content Policy Framework: Cross-Country Analysis

This section analyses the whole local content policy in its entirety across case study countries. The analysis facilitates the build-up of local content and how and why such measures of government intervention were undertaken by the different governments across the case study countries (Fig. 5.1).

5.5.1 Policy Statements

Policy statements are the governmental intentions and guiding principles that guide the whole local content policy within a given country. These guidelines are vast targets that might be directly or indirectly related to the oil and gas sector. They describe the pertinent areas in an oil and gas sector that the local content policy needs to attend to and create a mechanism for local integration in the oil and gas sector (Table 5.1).

Kenya, Uganda, Nigeria, Mozambique, Angola, Norway and Brazil have also adopted similar policy statements as a means of showing the

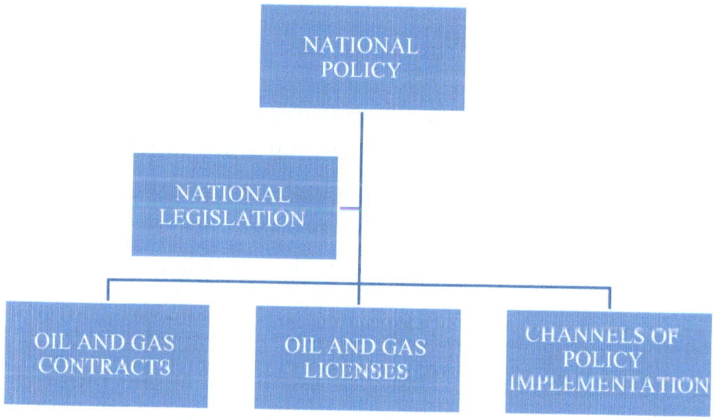

Fig. 5.1 The schematic presentation of local content policy build-up within a regulatory regime (*Source* Author's elaboration)

Table 5.1 Tanzania's policy statement on local content

Increase in exploration activities and discovery of natural gas both on-shore and offshore has attracted public attention on the potential benefits that are likely to accrue from oil and gas-related activities. To maximise such benefits, the Government recognizes the importance of Tanzanians' participation in the entire oil and gas value chain. This policy is formulated to provide a framework for Tanzanians participation in the oil and gas industry.

Source Tanzania's National Energy Policy, 2015. http://www.ewura.go.tz/wp-content/uploads/2016/08/National-Energy-Policy.pdf

Table 5.2 Brazil-Legislation for exploration and production of oil and natural gas

Tender Protocol: Article 15: Tender protocol shall be accompanied by the respective basic draft contract and shall indicate:
VIII- Minimum local content and other criteria related to the development of the national industry

Source Brazil Petroleum Exploration Law of 2010

legislator's intention towards the oil and gas industry. Oil and gas countries pursue local content obligations through a licensing regime. The licensing regime uses local content obligations as a bid factor for oil and gas companies to commit the local content obligations.

5.5.2 Licensing Regime

To depict the main criteria in giving awards, the Brazilian model in licensing round in 2013, the international oil companies placed their commitment in the Brazilian value chain. The responsibilities placed were 39 per cent for exploration, 62 per cent for the development of offshore acreage, and 85 per cent for the development of onshore acreage (Table 5.2).

A model licensing regime will have a commendable local content regime at the bidding stage. Firstly, it ensures that the bidding entities have placed a commitment towards specific industry benchmarks. The commitments set by international oil and gas companies creates a path for smooth Petroleum Agreement (PA) negotiations.

Table 5.3 Nigerian Oil and Gas Industry Content Development Act Section

3(1): Nigerian independent operators shall be given first consideration in the award of oil blocks, oil field licences, oil lifting licences and in all projects for which contract is to be awarded in the Nigerian oil and gas industry subject to the fulfilment of such conditions as may be specified by the Minister

Source Nigerian Oil and Gas Industry Content Development Act, 2010 Act No. 2

5.5.3 Primary Legislation

The primary legislation in the legal framework (usually the Hydrocarbon or Exploration and Production Law) is the significant part where the local content policy objectives envisaged in the licensing regime through the purport of the policy statements are enforceable and imposed on the international oil companies. In nascent oil and gas jurisdictions such as Kenya, Tanzania, Mozambique and Uganda without a preceding local content law. The local content definition is found in this Hydrocarbon Law or Exploration and Production Law. The implementation of local content into law varies from country to country according to their peculiarities. The extent to which these laws are soft or hard is dependent on the intention of the legislator (Table 5.3).

Local content obligations are not only found in primary legislation, and the legal framework can extend its application of local content obligations to secondary legislation. Secondary legislation will be in the form of regulations (Table 5.4).

Regulations in local content regimes also stipulate what regulatory authorities can regulate the local content implementation process within a given oil and gas industry. Regulations are the details to which extent the local content objectives are to be met, the limitations and the opportunities for both the local and foreign investors thereof.

5.5.4 Petroleum Agreements

Petroleum Agreements also have binding local content obligations. During the negotiations between the oil and gas investors and the

Table 5.4 Section 9: Priority of Ugandan goods and services during procurement

(1): Every licensee, contractors and subcontractor shall give preference to goods and services which are produced and are available in Uganda and services which are rendered by Ugandan citizens and companies
(2): Where goods and services required by the licensee or contractor or subcontractor are not available in Uganda, the goods and services, shall be provided by a company which has entered into a joint venture with a Ugandan Company
(3): The Ugandan company referred to under sub-regulation shall have a participating interest of at least forty-eight per cent in the joint venture

Source The Petroleum (Exploration, Development And Production) (National Content) Regulations, 2016

Table 5.5 Republic of Kenya model production sharing contract

Part IV- This part of the Petroleum Sharing Agreement in Kenya, advocates for employment and training provisions of Kenyans, the training fund that will be allocated to educating Kenyans, the preference of Kenyan goods and services, and provisions for technology transfer.

Source Schedule Republic of Kenya Model Production Sharing Contract

government. The Petroleum Agreements will provide an enhanced commitment with regards to safeguarding the expectations of both the oil and gas investors and the State. The Petroleum Agreements are used as vessels to ensure that there is a contractual relationship between the State and the oil and gas investors with regard to local content obligations (Table 5.5).

5.6 Initial Local Content Developments in Case Study Countries Country-by-Country Analysis

5.6.1 Norway

In Norway after the discovery of oil and gas resources in the Continental Shelf. The country introduced the Royal Decree of 1065 and 1972, respectively, to govern the development of the oil and gas industry.[17] The Norwegian government was keen to govern the oil and gas industry based on principle. Though in its early declarations the Decrees did not directly envisage local content policies, Article 4 guaranteed an exploration license to any local company, person or association.[18] Article 4 would not meet the demands of the local Norwegian. The Norwegian parliament introduced "the Ten Oil Commandments". These commandments were taken as principles governing the oil and gas industry in Norway. They later led to the development of local content policies to make sure that the oil resources were used to the benefit of the whole Norwegian society. The Local Content developments were first brought under Section 35 of the Resolution.[19] This provision made sure that Norwegian goods and services were given preference if they were competitive in both price and quality.[20]

The Ministry further created "the goods and service office" to ensure that Section 35 of the Resolution is administered effectively. It was only after acquiring 50–70% of the goods and services supplied, did the National entities such as Statoil, Hydro and Saga move to take up more tasks within the industry to become dominant players in their industry.[21] Norway did not only place attention to the international competition but also leveraged their local offshore shipping competence. They developed it to international standards and later created joint ventures through "State Participation Agreement", that later became a pivotal component for developing local content in Norway.[22]

---

[17] Royal Decree of 9 April 1965 relating to Exploration for and Exploitation of Petroleum Deposits in the Sea-Bed and its Subsoil on the Norwegian Continental Shelf.

[18] Hunter, T., 2014. Law and Policy Frameworks for Local Content in the Development of Petroleum Resources: Norwegian and Australian Perspectives on Cross-Sectoral Linkages and Economic Diversification, 27 MINERAL ECONOMICS 115.

[19] Holden, S., 2011 June, Avoiding the Resource Curse the Case Norway Department of Economics, University of Oslo.

[20] Ibid.

[21] Hunter, T., 2010. *Legal Regulatory Framework for the Sustainable Extraction of Australian Offshore Petroleum Resources. A Critical Functional Analysis*. University of Bergen.

[22] Ibid.

5.6.2 Nigeria

Nigeria implemented the Minerals Oil Act of 1914; the Act had local content provisions that were never realised into reality.[23] The Act envisaged that within ten years of the license being granted, the oil investors must have at least 75% local Nigerians in all tiers of the industry.[24]

Later Nigeria had the Petroleum Act (Drilling and Production) Regulations of 1969 that provided for local content provisions.[25] The Act governed the recruitment and training of local Nigerians.

To create synergy between the skills acquired due to the Petroleum Act. The Nigerian authorities enacted secondary legislation known as the Petroleum Training Institute (PTI) Act of 1972.[26] The sole purpose of the Act was to address and promote courses of instruction, training and research in the oil and gas sector. The Petroleum Training Institute could not keep up with the versatile oil and gas industry.[27] The skills undertaken at the time always fell short of the oil and gas best international practise. Nigeria further created the Petroleum Technology and Development Fund (PTDF) Act of 1973.[28] This Act was implemented to support and give more purpose to the PTI Act. The Fund was availed for training local Nigerians to facilitate their qualification as professionals, technicians and artisans in the fields of engineering, geology, science and management within the industry. The fund devastatingly failed in its implementation due to the misuse of funds by the elite of Nigeria.[29] Investigations showed that the fund was used for

---

[23] Raji, A.O.Y. and Abejide, T.S., 2014. The British Mining and Oil Regulations in Colonial Nigeria c. 1914–1960s: An Assessment. *Singaporean Journal of Business, Economics and Management Studies*, 51(1447), pp. 1–14.

[24] Ibid.

[25] Para. 38 (a) (i) of the First Schedule to the Petroleum Act.

[26] Petroleum Training Institutes Act.

[27] Heum, P., 2008. Local Content Development: Experiences from Oil and Gas Activities in Norway.

[28] Adekalu, S.O., Oludeyi, O.S., Genty, K.I. and Wolo, A., 2013. Petroleum Technology Development Fund (PTDF) Mandates and Human Capacity Development in Nigeria: Benefits for Nigerian Youths. *International Journal of Research in Management Issue*, 3, pp. 67–77.

[29] Ramírez-Cendrero, J.M. and Wirth, E., 2016. Is the Norwegian Model Exportable to Combat Dutch Disease?. *Resources Policy*, 48, pp. 85–96.

addressing "pressing national interests" which violated the law.[30] In the late 1980s, Nigeria emphasised the indigenisation policy or the "Nigerianization" of the oil and gas industry in Nigeria.[31] The policy aimed to retain ownership and control of the oil and gas resources, and this policy lasted from 1989 to 2000. However, the PTI Act and PTDF Act had already increased local participation. Very little transfer of technology and "Nigerianization" had taken place in this regard. The Indigenisation policy was a vast radical move by the Nigerian authorities aimed at increasing local participation in the oil and gas industry. By the year 2000, the Christopher Kolade inquiry had discovered irregularities, such as the high levels of corruption and lack of transparency.[32] That led to 31 licenses being revoked that were previously awarded through the indigenisation policy.[33]

5.6.3 Brazil

President Lula de Silva held that the development of the oil and gas sector must be seen from a holistic approach.[34] The national oil company Petrobras was to be seen as a vehicle for development that reflects Brazilian interests. Due to such an approach, the government resorted to "increase and maximise the share of domestic industry participation in supplying goods and services to the oil and gas sector".[35] Local content policy in Brazil was initially seen as a process of selecting winners. This was attached with a signature bonus and a minimum exploratory programme. The government expressed their intention in making the Brazilian oil and gas industry a local industry. This was done by having local content requirements as an integral part when bidding for licenses

[30] Ibid.

[31] Ogbuagu, C.S., 1983. The Nigerian Indigenization Policy: Nationalism or Pragmatism? *African Affairs*, pp. 241–266.

[32] Nwaobi, G.C., 2004. Corruption and Bribery in the Nigerian Economy. Available at SSRN 506342.

[33] Ibid.

[34] Lima de Oliveira, R., 2016. Resource-Led Industrial Development in the Oil & Gas Supply Chain: The Case of Brazil.

[35] Ibid.

in Brazil in rounds 5 and 6 (2003–2004).[36] But from round 7 onwards, local content requirements became mandatory by issuing a certification system.[37] Where items and sub-items are to be measured according to their degrees of value addition as accredited by ANP. An official ANP booklet designates the items and sub-items and the minimum local content commitment and its value thereof. Penalties were issued for not complying.[38] The creation of a booklet and a certification system was too accurate of a metric used in estimating the actual investments in local content. The controversy is that ANP set the domestic content to be added into the Brazilian economy. The ANP obtained its information from entrepreneurial organisations such as Brazilian Machinery and Equipment Association (ABIMAQ).[39] Such organisations looked for higher targets to accrue benefits for its members. The National Oil and Natural Gas Industry Mobilization programme (Prominp) was established to facilitate the growth of the oil and sector with the sole purpose of meeting the strict local content requirements.[40] Prominp coordinates between different main stakeholders of the oil and gas industry to identify the industrial bottlenecks and develop measures that can enhance the local capacity and competitive advantage. One of the critical sectors incentivised by Prominp and other government initiatives was the shipbuilding industry.[41] Brazil's competence in the shipbuilding industry was fully leveraged for the production of tankers and other vessels for use in offshore oil production.

[36] Tordo, S., Warner, M., Manzano, O. and Anouti, Y., 2013. *Local Content Policies in the Oil and Gas Sector*. The World Bank.

[37] Lima de Oliveira, R., 2016. Resource-Led Industrial Development in the Oil & Gas Supply Chain: The Case of Brazil.

[38] Filho, N.N. Brazil's Oil & Gas Local Content Policy: Lessons Learned. Institute of the Americas.

[39] Filho, N.N. Brazil's Oil & Gas Local Content Policy: Lessons Learned. Institute of the Americas.

[40] Cavalheiro, G.M. do C., Joia, L.A., and Gonzalves, A.C., 2014. Strategic Patenting in the Upstream Oil and Gas Industry: Assessing the Impact of the Pre-Salt Discovery on Patent Applications in Brazil. *World Patent Information*, 39, 58–68.

[41] Lima de Oliveira, R., 2016. Resource-Led Industrial Development in the Oil & Gas Supply Chain: The Case of Brazil.

5.6.4 East Africa

Local content policies in East Africa have evolved in the same dimension as the previous oil and gas mature countries such as Norway, Brazil, Nigeria and Angola. The recent discoveries of huge oil and gas reserves in Mozambique, Tanzania, Uganda and Kenya have recently attracted large oil and gas investments in East Africa. The countries possess significant endowments in other natural resources attracting large-scale investments in their extractive sector. Kenya discovered over 700 million barrels of oil in Turkana. Uganda discovered about 6.5 billion barrels of oil in the Albertine region. Tanzania has found 57 trillion cubic feet of proven gas reserves in the Southern part of Tanzania and Mozambique discovered gas in the Rovuma Basin worth 425 billion cubic metres of gas.[42] With these massive discoveries in East Africa, the region makes its mark on the rich extractive countries in the region. The main hurdle is transforming these discoveries into economic prosperity for the general population. The remote regions in East Africa have been blessed with these oil and gas resources. The local communities in these regions have been marginalised. They are therefore troubled and unstable due to the high rates of poverty, high illiteracy rates, and rampant spread of HIV as well as poor delivery of social services.[43] East African countries started their oil and gas industries without having an expressive and illustrative local content regime. Tanzania was probably the only country that had the developed local content legislation, which was found in the Petroleum Act, 2015. Kenya, Mozambique and Uganda did not have local content legislation in the initial steps of developing their oil and gas resources. But without delay, they followed suit. The East African countries look to other oil and gas-rich countries where the local content policies have been implemented.

[42]Hansen, M.W., Buur, L., Mette Kjær, A. and Therkildsen, O., 2016, May. The Economics and Politics of Local Content in African Extractives: Lessons from Tanzania, Uganda and Mozambique. In *Forum for Development Studies*, 43(2), pp. 201–228. Routledge.

[43]Chrispine Odour et al., 2014. The Political Economy of the Extractive Sector. In Miriam, W. Omolo, O. and Mwabu, G., eds., *A Primer to the Emerging Extractive Sector in Kenya: Resource Bliss, Dilemma or Curse* (p. 181). Nairobi: Institute of Economic Affairs.

5.6.5 The Gulf of Guinea

After the discovery of oil and gas in the Gulf of Guinea. A consensus was reached as to the state's crucial position in leveraging the natural resources for the benefit of the local population. Subsequently, state institutions were formed with different functional responsibilities for policymaking, technical management, and trade participation. Following the measures taken by Nigeria and Angola, the Local Content Policy Framework of Ghana has the primary policy objective of "maximising the benefits of oil and gas wealth generation on an integrated local content platform by enhancing the use of local skills, products and services, generating employment for people, businesses and financing in all aspects of the value chain of the oil and gas industry, and ensuring value addition in Ghana".[44] In Liberia, the 2012 National Petroleum Policy sets the governance structure for the sector. That covers main aspects of natural resource management such as resource ownership and maritime borders, legal system and institutional supervision, licencing, state involvement, revenue management, health, safety, and climate and local content.[45] Equatorial Guinea has introduced a local content strategy to pressure foreign oil companies and service providers to hire locals and to buy local products and services. The Ministry Order 1/2014, for example, establishes the policy basis for industry regulation in Equatorial Guinea.[46] This order requires all petroleum contracts to have provisions on local content tailored towards job growth, capacity development and participation by the national industry.

[44]Obiri K.A., Bjeirmi, B. and Boateng, P. 2019. Local Content Implementation Enhancement Through Infrastructure Development in Ghana's Oil and Gas Industry. *Journal of Energy Research and Reviews*, pp. 1–10.

[45]Klueh, U.H., Pastor, G. and Segura, A. 2009. Policies to Improve the Local Impact from Hydrocarbon Extraction: Observations on West Africa and Possible Lessons for Central Asia. *Energy Policy*, 37(3), pp. 1128–1144.

[46]Oyewole, B., 2018. Overview of Local Content Regulatory Frameworks in Selected ECCAS Countries.

5.6.6 Local Content Institutional Framework

Local content has fundamentally been regulated by a pyramid of institutional capacities and functionaries within a given country. The Institutional framework is created by legislation that governs and promulgates local content policy into the oil and gas industry.

Most governments have the institutional framework based on policymaking function, a technical control function and a political and commercial function. Different countries have different institutional frameworks; traditionally, the institutional framework includes the Energy Ministry, a Local Content Committee, a national oil company/or a local content parastatal.

In Norway, the Parliament approved the creation of crucial oil and gas institutions namely the Ministry of Petroleum and Energy, the Norwegian Petroleum Directorate and the national oil company Statoil (Now Equinor).[47] In Brazil Conselho Nacional de Politica Energetica creates guidelines for local content policies in harmony with the Ministry of Mines and Energy. The regulatory activities fall under the Agencia Nacional do Petroleo. Banco Nacional de Desenvolvimento Economico e Socialal also sets minimum local content requirements for local content initiatives financed by them. Several independent institutions, of which National Organization of the Petroleum Industry (ONIP) is the largest, carry out vendor's certification programmes, training and advocacy activities.[48]

In Nigeria, the Presidency has very close ties to the oil and gas industry. The Ministry of Petroleum has the overall authority to create institutions that regulate local content in the oil and gas sector. The Nigerian National Petroleum Corporation (NNPC), the national corporation was created to bring about capacity in the Nigerian market.[49] The Department of Petroleum Resources established as petroleum as well as

[47]Heum, P., 2008. Local Content Development: Experiences from Oil and Gas Activities in Norway.

[48]Tordo, S., Warner, M., Manzano, O. and Anouti, Y., 2013. *Local Content Policies in the Oil and Gas Sector*. The World Bank.

[49]Balouga, J., 2012. Nigerian Local Content: Challenges and Prospects. *International Association for Energy Economics*, 4, pp. 23–26.

the governmental body with the mandate of regulating local content in Nigeria. Through the Nigerian Local Content Development Policy, Nigeria created the Nigerian Content Development Monitoring Board (NCDMB).[50] This Board entrenches the planning and contracting process of local Nigerian firms into the oil and gas value chain. Tanzania has an array of institutions when it comes to local content policy in the oil and gas industry. These institutions have overlapping authorities when regulating oil and gas activities in Tanzania. The national oil company Tanzania Petroleum Corporation (TPDC) has the mandate to ensure that local content provisions are adhered to when concluding contracts with multinational corporations.[51] The Ministry of Petroleum is the overall authority in regulating oil and gas activities in Tanzania. Within the Ministry, there is an upstream regulatory authority, The Petroleum Upstream Regulatory Authority (PURA).[52] This authority ensures that there is local participation in the upstream oil and gas activities.

Furthermore, the government created the National Economic Empowerment Council (NEEC), was designed to have sole custody in local content oil and gas regulation in the oil and gas industry. The Council is under the Prime Minister's Office that takes the lead on local content issues and has appointed focal points within each ministry and parastatal to coordinate local content promotion.[53] The Tanzania National Business Council (TNBC) and the UONGOZI Institute. These are all part of governmental bodies created to be involved in research and training for government officials on different aspects of oil and gas industry such as local content.[54] In Kenya, there is also an array of institutions when it comes to regulating oil and gas operations. The National Construction Authority ensures there is local participation

[50] Ibid.

[51] Ovadia, J.S., 2017. Local Content in Tanzania's Gas and Minerals Sectors: Who Regulates? CMI Brief.

[52] Ibid.

[53] Lee, B. and Dupuy, K., 2018. Understanding the Lie of the Land: An Institutional Analysis of Petro-Governance in Tanzania. *Journal of Energy & Natural Resources Law*, 36(1), pp. 85–101.

[54] Ibid.

in the construction, roadwork, and waterworks electrical and technical faculties in the creation of oil and gas tools in Kenya.[55]

The Benefit Sharing Authority ensures the gains of the policy are distributed to everyone and each one in Kenya. The Local Content Development Committee has the mandate to develop and follow up on the local content policy requirements and the Local Content Development and Monitoring Unit. This body monitors local content development in Kenya.[56] In Uganda, The Petroleum Exploration, Development and Production Act (PEDP Act, 2013) and the Petroleum Refining, Conversion, Transmission and Midstream Storage Act (PRCTMS Act, 2013) create the current institutional and regulatory framework for the governance of Uganda's oil and gas sector (especially local content. The Directorate of Petroleum under the Ministry of Energy and Mineral Development (MEMD) is responsible for local content policy creation as well as ensuring that the policy commitments are entrenched in the licensing of the oil and gas activities.[57] The Petroleum Authority of Uganda (PAU), established in 2013, is the regulatory body responsible for monitoring local content compliance in the oil and gas sector. Finally, the Uganda National Oil Company was established (also in 2013) to manage the state's commercial interests in the oil and gas sector as well as ensure local participation in Ugandan oil and gas value chain. Both the MEMD Directorate and PAU consist of "National Content Units", with the former leading on issues related to policy formulation, training, and enterprise development. In contrast, the latter fundamentally focuses on monitoring compliance of the local content policy commitments. In the Gulf of Guinea, local content policies are implemented radically and monitored by The Ministry of Petroleum is responsible for: (1) the formulation of local content policies, (2) the approval of foreign workers recruitment and annual recruitment plans, and (3) approval of capacity

[55]Murungi, C.N. Introducing Local Content Regulation in Kenya's Economy: A Case Study of Kenya's Energy, Oil and Gas Sector. Institute of Diplomacy and International Studies. University of Nairobi. R50/63843/2010.

[56]Ovadia, J.S., 2016. Local Content Policies and Petro-Development in Sub-Saharan Africa: A Comparative Analysis. *Resources Policy*, 49, pp. 20–30.

[57]Sen, R., 2018. Enhancing Local Content in Uganda's Oil and Gas Industry (No. 2018/110). WIDER Working Paper.

building plans for nationals. The Ministry or a parastatal normally publishes a list of domestic goods suppliers that foreign investors can procure from with capacity to supply the oil and gas industry. In Equatorial Guinea, Ghana, Liberia, Angola, these countries have a fund that finances the local content initiatives in their economies. With a technical and vocational institution that incubates the technology and R&D transfer from oil and gas investors to the nationals. The national oil and gas companies established in the Gulf of Guinea such as GEPetrol of Equatorial Guinea, SONANGOL of Angola, the National Oil Company of Liberia, and the Ghana National Petroleum Corporation-they all negotiate a local content provision in the various petroleum agreements, managing licensing rounds and managing the procurement of goods and services. The national oil companies are the primary vehicle of driving a country's interests in the oil and gas industry. In turn means that national oil companies have the mandate of cracking through the enclave of the oil and gas industry using the right procedures, recognising the different interests of various stakeholders, ensuring an equitable distribution of the benefits form the oil and gas industry to the general population.

5.6.7 An Illustration of Oil and Gas Sectoral Linkages

The discovery of oil and gas resources within a country is a double-edged sword, meaning that the country can either slip away in disarray with extractive commodities causing massive inflation through rapid currency appreciation, leading to what is known as the Dutch Disease.

On the other hand, oil and gas resources can be used as a vehicle towards economic prosperity and independence for a resource-rich country. Factors such as the previous economy of the country before discovering oil and gas resources, the institutional and regulatory frameworks created as well as the principles of good governance are all part of the pile making oil and gas resources lead to a commendable economic outcome. Although all these mentioned aspects are vital in that transplantation process of oil and gas resources into economic development. The ability of a country to develop sustainable linkages between economic sectors to the other is also essential. These linkages have spill-over effects onto an economy. In countries like Norway and Brazil after the discovery of oil, there was an increment in tax revenues (fiscal linkages). This created possibilities for the respective public sectors to expand R&D and higher education to sustain the high levels of regional economic development and to increase technological development initiatives.[58] In Nigeria the amounts gained from the servicing the oil and gas industry such as fabrication, engineering, procurement, construction (EPC), Front End Engineering Design (FEED), conceptual designs and seismic studies.[59] The proceeds of these transactions were taken out of the country to other manufacturing countries of the goods and tools utilised to service the Nigerian oil and gas industry.[60] In Tanzania, Kenya and Uganda, FDI offers a way out towards economic prosperity and a means to be free from donor dependence. The question arises whether the oil and gas FDI leaves lasting benefits in these countries through linkages that award a fair and equitable distribution within the local oil and gas industry. The existing legal frameworks in Tanzania, Kenya and Uganda do not create means for or advocate for linkages from the oil and gas sector to other sectors. Hence the creation of linkages is left to the market forces. In these countries, the oil and gas industry is a nascent industry that is heavily politicised due to the industry's economic potential and its political implications.[61]

The Norwegian and Brazilian governments were able to leverage their oil and gas resources to achieve significant improvements in their skills and R&D. This led to concluding technology agreements at the beginning of the oil and gas industry to encourage foreign oil companies to

---

[58] Intergovernmental Forum on Mining, Minerals, Metals and Sustainable Development 2015. Norway. Horizontal Linkages. Local Context as the Main Driver for Horizontal Linkages.

[59] Ihua, U.B., Ajayi, C. and Eloji, K.N., 2009, May. Nigerian Content Policy in the Oil and Gas Industry: Implications for Small and Medium-sized Oil Services Companies. In Proceedings of the 10th Annual Conference, IAABD.

[60] Ihua, U.B., Ajayi, C. and Eloji, K.N., 2009, May. Nigerian Content Policy in the Oil and Gas Industry: Implications for Small and Medium-sized Oil Services Companies. In Proceedings of the 10th Annual Conference, IAABD.

[61] Hansen, M.W., 2013. Reaping the Rewards of Foreign Direct Investment: Linkages between Extractive Multinational Corporations and Local Firms in Tanzania.

invest in R&D in Brazil and Norway.[62] Both governments called for a collaborative approach between the international oil companies and the Universities that at least required 50 per cent of research to be conducted in Norway and Brazil.[63] Hence the development of the Norwegian and Brazilian oil and gas industry was done domestically within their borders thereof. This led to the creation of research centres, such as Sintef, Christian Michelsens Research, and Rogaland Research, in Norway. While in Brazil, they also created knowledge centres that coordinated with the Petrobras Research Center (CENPES).[64] Oliveira and Mendonca stated that "CENPES, using resources from Law 9.478/1997, underwent a recent expansion to occupy an area of 300,000 square meters, making it one of the biggest research complexes in the world".[65]

Establishing a policy that directs the cultivation and development of linkages led to Norway and Brazil exporting their offshore service expertise to the rest of the world industries (e.g. automation systems, detection and communication systems).

In Kenya, Tanzania and Uganda, these countries do not possess any type of policy to foster linkages. The prominent oil and gas investors do not have any local counterpart to match their industrial demands. This makes it challenging to implement the specific policy requirements of local content. Thus minimal linkages are developed. Institutions are a vital aspect when developing linkages through local content policies. Institutions create checks and balances that guide the direction of the policy. In Norway and Brazil institutions have been divided to meet the different demands of the local content policy. In Norway and Brazil, the governments separated business from politics through the creation of a national oil company. The separation was not done merely for ideological purposes but also practical and efficiency purposes. Asiago states that "while businesses were to operate in the realm of business principles to

[62]Noreng, Ø. 2009. Brazil and Norway–Offshore Petroleum Experiences and Lessons. The *Journal of Energy and Development*, 35(1/2), pp. 79–99.

[63]Noreng, Ø. 2009. Brazil and Norway–Offshore Petroleum Experiences and Lessons. *The Journal of Energy and Development*, 35(1/2), pp. 79–99.

[64]Mendonça, R.W. and de Oliveira, L.G., 2013. Local Content Policy in the Brazilian Oil and Gas Sectoral System of Innovation. *Latin American Business Review*, 14(3–4), pp. 271–287.

[65]Ibid.

create international competitive frameworks, it was no doubt that such principles were to be governed by transparent political decisions".[66] The institutional design in Norway and Brazil minimises the risk of having a conflict of interest since the NOCs, namely Petrobras and Equinor (formerly known as Statoil) only concentrated on the commercial activities. The creation and development of linkages are highly dependent on how institutions are efficient and transparent in the oil and gas industry. Oil and gas operations in Norway and Brazil have not been dependent on the quality of their leaders and their visions but o the quality of their institutions.[67]

Political settlement is also a huge factor in the oil and gas industry and to what extent linkages can have breadth and depth in an economy. In most Sub-Saharan countries, the political structure has explicitly used natural resources, particularly the oil and gas industry for the gain of the political elite. Kjær, Katusiimeh and Whitefield state that "the increasing fragmentation and inflationary patronage have both indirect and direct implications for the extent to which local linkages with the oil company can be nurtured. Indirectly, implementing industrial policy to promote production and build capacity for linkage formation meets a larger number of obstacles and resistance when important groups must be accommodated. Productive sectors are in general only promoted if they are important to the ruling coalition".[68] Norway and Brazil have the available skills in their economy that can create the absorptive capacity. In Nigeria, the economy lacks adequate skills that can foster linkages, and this made their development limited thereof. Nascent oil and gas Sub-Saharan economies have not addressed knowledge gaps in the creation of their local content policies. This conclusively cannot lead to policy creation on linkage formation within their economies.

[66]Asiago, B.C., 2017. Norwegian Local Content Model a Viable Solution. US-China Law Review, 14, p. 471.

[67]Norway at the forefront of extractive industries transparency efforts.

[68]Kjær, A.M. and Katusiimeh, M., 2012. Growing But Not Transforming: Fragmented Ruling Coalitions and Economic Developments in Uganda (No. 2012: 07). DIIS Working Paper.

5.7 Examples of Energy Justice and Local Content in Sub-Saharan African Countries

Oil and gas resources are discovered in marginalised areas and areas that have poor oil and gas resource governance skills and mechanisms. In Sub-Saharan Africa, the oil and gas resources have been found in remote rural areas that are populated by local people, and these people have been regarded as unproductive, with fragile eco-systems and high insecurity.[69]

In all case study countries, namely Norway, Brazil, Nigeria, Kenya, Tanzania and Uganda, have all found oil and gas resources within the vicinity of local people in these countries. The Southern Tanzania local people in Mtwara and Lindi, the Saami people in Northern Norway, the local territories of Brazil the lands of the Suruwaha, Banawa, Deni and Paumari Indians, the local people of the Niger Delta, the local community in the Albertine Region of Uganda and lastly the local people of Turkana in Kenya. The oil and gas industry affects different spheres of a community, society and country. Namely the economic, political, and sociocultural aspects at different levels within the oil and gas industry. These ramifications that call for the advocacy of justice, particularly energy justice as a constant factor in policy adaptation and implementation.

At the macro-level, oil and gas resources have fundamentally not been exemplary due to the unequal distribution of wealth between the local people of a given country.[70] This has led to political instability in numerous oil and gas-rich countries such as Nigeria, Angola, Cameroon, Chad, Kenya and Tanzania, to name a few.

These local people suffer from dispossession from land and degradation of the environment due to the oil and gas industry operations at the micro-level where the local people who have the oil and

[69] Mkutu, K., Mkutu, T., Marani, M. and Ekitela, A.L., 2019. New Oil Developments in a Remote Area: Environmental Justice and Participation in Turkana, Kenya. *The Journal of Environment & Development*.

[70] Mkutu, K., Mkutu, T., Marani, M. and Ekitela, A.L., 2019. New Oil Developments in a Remote Area: Environmental Justice and Participation in Turkana, Kenya. *The Journal of Environment & Development*.

gas resources within their vicinity. This undermines the local people's economic ecosystem along with conflicts, tensions in the ethnic minorities.[71] Energy justice, in this case, is related to how local content policies facilitate the inclusion and participation of the local people into the oil and gas value chain. Energy justice determines to what extent local content policies disseminate equal inclusion and exclusion patterns with the critical aim of cracking through the enclave nature of the oil and gas industry. The concept of energy justice merges its discussions on the principles of justice, and that is considering the economic, social aspects and implications of energy resource development onto the immediate community.[72] The debate on energy justice principles being entrenched in the adaptation and implementation of local content policies must move beyond the mere distribution harms and benefits. The three tenets of energy justice, namely distributive justice, recognition and procedural justice must be incorporated into the legal and regulatory process which creates the local content policies within a given oil and gas-rich country.

5.7.1 Energy Justice Tenets and the Term "Local" in Local Content

The principles of energy justice can be used to discuss what the meaning of "local" in local content is ought to be. Most oil and gas countries (with little mention of Norway in this regard). Define local in their local content regimes through a centralist approach.[73] Defining "local" in local content from a centralist approach may ring-fence the benefits out of the local people's reach. The tenets of energy justice, such as justice as recognition, would not be prevalent in the centralist approach. Because not all people would have been recognised with this centralist approach.[74] The tenet of energy justice, distributional justice would also

[71] Barczewski, B., 2013. How Well Do Environmental Regulations Work in Kenya?: A Case Study of the Thika Highway Improvement Project. Center for Sustainable Urban Development.

[72] Schlosberg, D., and Carruthers, D. 2010. Localstruggles, Environmental Justice, and Community Capabilities. *Environmental Politics*, 10(4), pp. 12–35.

[73] Nwapi, C., 2015. Defining the "Local" in Local Content Requirements in the Oil and Gas and Mining Sectors in Developing Countries. *Law and Development Review*, 8(1), pp. 187–216.

[74] Ibid.

not be upheld in the centralist approach since there will be an uneven distribution of wealth between different local people of the same country. The collaboration of energy justice in defining "local" in local content provides a holistic approach-an approach that recognises the local populations. When there is recognition of the different local people, there is an adequate and fair distribution of harms and benefits, and the right procedures have been undertaken to recognise and distribute the harms and benefits of being blessed by oil and gas resources. It is automatic that there will be no resource conflicts, community frustration and most importantly, the inclusion of subnational stakeholders. In Nigeria, the Nigerian Oil and Gas Industry Development Act, 2010 defines Nigerian Content as "the quantum of composite value added to or created in the Nigerian economy by a systematic development of capacity and capabilities through the deliberate utilization of Nigerian human, material resources and services in the Nigerian oil and gas industry".[75]The Act advocates for "first consideration" should be given to goods and services manufactured within Nigeria as well as training and employment opportunities to be given to Nigerians in the first consideration.

The term "first consideration" has synergy with justice as recognition since the provision recognises the need to have economic and technical inclusion of Nigerian goods and services in the oil and gas industry value chain, regardless of social-cultural, ethnic, racial and gender differences thereof.[76] The Act does not give a proportion as to what content must be sourced from within the locality where the oil and gas operations take place. Section 27 of the Act, just allocates powers to the Board to require a company to have an office within the community having oil and gas operations.[77] The Act does not give "fist consideration" on who must be employed in the office, or what is the mandate or obligation the office. It is such provisions that lead to NIMBYISM (Not-in-my-backyard) tendencies and sentiments. This is because the local individuals feel that they are not part of the oil and gas value chain. They felt left out and robbed of what they own within their land and vicinity. This enhances

[75]Nigerian Oil and Gas Industry Content Development Act, 2010, s 106 [NOGICD Act].
[76]Fraser, N., 2009. Social Justice in the Age of Identity Politics. Geographic Thought: A Praxis Perspective, pp. 72–91.
[77]Nigerian Oil and Gas Industry Content Development Act, 2010, s 106 [NOGICD Act].

public resistance as seen in the Niger Delta through the presence of militants and infrastructural destruction due to the economic realities that given rising to a Nigerian elite vis-à-vis a Nigerian poverty-stricken indigenous.[78]

On the other hand, the Act defines a Nigerian company (same as defining a local company) as "a company formed and registered in Nigeria following the provision of Companies and Allied Matters Act with not less than 51% equity shares by Nigerians".[79] The definition of a Nigerian company falls short of the principles of energy justice in the Nigerian oil and gas value chain. In terms of local content policy, the steep requirement of having Nigerians own at least 51% equity for a company to be a Nigerian company does not promote distributional justice within the Nigerian oil and gas value chain. The Nigerian Content Act is exhaustive and illustrative compared to other Local Content Acts within the continent. But the high equity requirements make it hard for middle-class and poor Nigerians to be involved in their industry. This results in an unequitable distribution of benefits among local Nigerians.

The Act, entails there to be an office in the local community where the oil and gas operations take place. The presence of "just an office" falls short of justice as recognition. Since it is not demarcated in the Act as to who or what community is recognised to be part of the office, furthermore, it has not been established what procedure should be undertaken to be employed in the office. In Tanzania, the Local Content Policy for the Oil and Gas Industry defines local content as "the added value brought to the country in the activities of the oil and gas industry in the United Republic of Tanzania through the participation and development of local Tanzanians and local businesses through national labour, technology, goods, services, capital and research capability".[80] At another point, the policy defines local content as "the added value brought to a host nation (and regional and local areas in that country) through the activities of the oil and gas industry".[81] The reference made to "regional

[78]Okoli, A.C., 2019. Oil Pipeline Vandalism in the Niger Delta: Need, Greed and Grievance Factors.

[79]Nigerian Oil and Gas Industry Content Development Act, 2010, s 106 [NOGICD Act].

[80]United Republic of Tanzania, Local Content Policy, 2014.

[81]United Republic of Tanzania, Local Content Policy, 2014.

and local areas" shows that there is recognition of the local people where the oil and gas resources are being extracted. Energy justice does not recognise the local people but also the procedure and the means that are to be used to distribute the benefits and harms among the local people of Tanzania. "Local" is defined under the policy as "the Tanzania Mainland and its people". This meant that there must be a preference given to Tanzanian nationals.

But the mention of Tanzania Mainland negates the Tanzanians in the Unguja and Pemba Islands where they may have oil and gas resources. The policy has a "deliberate preference" to be given to Tanzanians in employment as well as goods and services[82]—the policy advocates for both community content and national content in the same policy. The dual presence of both community content and national content in the same policy draft makes local content futile in its application as the policy does not have precision on whom it addresses.

Energy justice throughout the local content policy has to evaluate where injustices emerge, which affected section of the community is to be addressed, which processes exist for their remediation to reveal and reduce the injustices.[83] Having a community and national content policy at the same time within the same local content policy does not evaluate the injustices of the local people not being involved in their oil and gas value chain. Since the policy advocates for both the immediate community and the nation at large, it is difficult to establish which community is most affected by the inadequacies of the policy. These inadequacies do not avail the remedy to reveal injustices of the oil and gas industry. Uganda and Kenya have both adopted the same national content policy. Kenya and Uganda, Mozambique, Equatorial Guinea and Liberia did not have an elaborative and illustrative policy as Tanzania at the time of the discovery of the oil and gas resources. In Uganda the national content is defined as the "means the quantum or percentage of locally produced materials, personnel, financing, goods and services utilized by a person, body or entity subject to this Act in any operation, undertaking

[82]Ibid.

[83]Jenkins, K., McCauley, D., Heffron, R., Stephan, H. and Rehner, R., 2016. Energy Justice: A Conceptual Review. *Energy Research & Social Science*, 11, pp. 174–182.

or activity carried in Uganda and which can be measured in monetary terms."[84]

While Kenya;

In Kenya, Local content is defined as "the means of maximising the level of use of local goods and services, people, business and financing. While local participation means "maximising the depth and breadth of local ownership, control and financing to increase local value and capture from all parts of the value chain created from the resource including those activities in which nationals, local business and capital are not currently engaged at home and abroad".[85]

Both Uganda and Kenya don't mention host communities in their application of local content policies. The local in the Albertine region in Uganda and the local people of the Turkana have all been politically and economically excluded from different industries before the discoveries of oil and gas resources. It is undesirable that the local content policies in these two regimes are being used to provide a social license to operate rather than justice through local integration in the oil and gas value chain. In Brazil and Norway, the approach towards defining local content took a rather unconventional method. In Brazil, local content means "the proportion between the value of the goods produced and services rendered in Brazil following the terms of the Model Contract 2015 and the total value of the goods used and services provided for this purpose".[86] Norway did not have a definition for local content but rather adopted "the ten commandments of the oil and gas industry". The adoption of "the ten oil commandments" and the emphasis on value rather than prescribed metrics in the Brazilian and Norwegian definition are among the many reasons as to why the Brazilian and Norwegian local content policies have been among the most successful local content regimes to date. In these local content regimes, there is room for flexibility to the ever-changing demands of the oil and gas industry. The implementation of the policy objectives reaches the demands of energy justice within a local content regime.

[84]Local Content Bill, 2017.

[85]Section 2, Local Content and Participation Bill.

[86]Brazil Model Contract 2015, Art.1.3.15.

5.7.2 Energy Justice and Local Content Requirements

Local content requirements in oil and gas industries have different quotas for different countries, and they are progressive and regressive dependent on the maturity of a given industry in a particular country. In Nigeria, the percentage given as a requirement for a local company to qualify as a local company is a company that has not less than 51% equity shares by Nigerians.[87] In Uganda, the local company quota is that Ugandan nationals must own t least 48% of the equity shares for a company to qualify as a local company.[88] In Tanzania, the percentage prescribed for a company to be eligible as a local company was 51% but later was amended to 20%.[89] In Kenya, the prerequisite for a local company to qualify as a local company. The company must be "effectively owned and controlled by nationals of Kenya".[90] In Brazil and Norway, on the other hand, the local content regimes do not prescribe a given quota that must be met for a local company to qualify as a local company. The most crucial aspect to be taken into account is that the oil and gas companies do business in Norway or Brazil. What the Norwegian and Brazilian local content regimes seek is not the rules and regulations but the value that is to be added in the local community as well as create linkages within different industries thereof. For there to be energy justice in local content requirements the underlying objective and rationale for formulating a given local content regime must reflect the culture, historical perspective, socio-political considerations as well as economic realities of the given country. Emphasising on the "objectivity and rationality" of a local content policies meets the demands of energy justice.

Energy justice aims to distribute the goods of local content policies while recognising all relevant communities or people following a just due process. This means that the local content requirements prescribed must be restorative. This is because local content policies are enacted as a

[87]Nigerian Oil and Gas Industry Content Development Act, 2010.

[88]The Petroleum(Exploration, Development and Production) Act, 2013.

[89]Mining (Local Content)(Amendments) Regulations, 2019.

[90]Local Content Bill of Kenya, 2018.

restorative tool, which can integrate the previously marginalised communities through law and policy. Heffron and McCauley state that "if restorative justice were applied to the energy sector it would ensure that decision-making was made in light of considering the potential harm of that decision and consequently the true cost of that decision".[91] Strict local content requirements prescribed by these countries have not taken into account the realities of their economies.

Furthermore, mature local content regimes such as Nigeria in this case study have depicted the results of creating a local content regime that does not reflect the economic realities of an economy. For nascent Sub-Sahara African countries to have an efficient local content policy. The local content requirements must be made in light of their guilt of causing harm by marginalising the vulnerable groups within the economy. It is only in this way would the local content regimes meet the demands and principles of energy justice.

5.8 Summary

The oil and gas industry is meant to propel economies into prosperity. The leverage of having these resources that attract a Gold Rush to the remote areas of any country is significant for any country's growth. But linkage development from the developed sectors to the low productive sectors of the economy are paramount for there to be any growth. But the oil and gas industry is a capital-intensive industry that requires sophisticated skills, well-developed infrastructure, high levels of technology and sound natural resource governance for there to be any tangible developments from this sector. Mature oil and gas economies have been able to achieve a probable outcome by designing and adopting efficient local content policies. That understand the market forces, avoiding being nationalistic to avoid investor-flight. As well as develop suitable natural resource governance mechanisms that are inclusive and versatile according to global markets. Countries such as Norway and Brazil

[91] Heffron, R.J. and McCauley, D., 2017. The Concept of Energy Justice Across the Disciplines. *Energy Policy*, 105, pp. 658–667.

have been the poster boys for exemplary local content policies. One for their flexibility and rigidity, two for their ability to address market demands and develop monopolies through their national oil companies. Sub-Saharan oil and gas heavy weigh producers Nigeria and Angola. Have tried to replicate the Norwegian/Brazilian model with no success. But challenges in areas of inadequate domestic skills, poor infrastructure, lack of technical and financial capacity of domestic firms and corruption must be addressed. The issues of regulatory capture of local content watchdogs in Sub-Saharan Africa must also be addressed. Having natural resources in itself has proven over the years not to be the primary vehicle to economic prosperity. This is attributed to the enclave nature of the oil and gas industry. Energy justice provides a tool of cracking through this enclave, as shown in this Chapter, though Norway and Brazil did not explicitly name justice in their local content provisions. But the inclusivity, due process and recognition of various stakeholders regardless of the national oil company monopoly shows that having a "just" local content framework is the critical aspect in cracking the oil and gas enclave.

References

Adekalu, S.O., Oludeyi, O.S., Genty, K.I. And Wolo, A., 2013. Petroleum Technology Development Fund (PTDF) Mandates and Human Capacity Development in Nigeria: Benefits for Nigerian Youths. *International Journal of Research in Management Issue*, 3, pp. 67–77.

Andersen, S.S., 1993. The Struggle Over North Sea Oil and Gas: Government Strategies in Denmark, Britain, and Norway. A Scandinavian University Press Publication.

Asiago, B.C., 2017. Norwegian Local Content Model a Viable Solution. *US-China Law Review*, 14, p. 471.

Balouga, J., 2012. Nigerian Local Content: Challenges and Prospects. *International Association for Energy Economics*, 4, pp. 23–26.

Barczewski, B., 2013. How Well Do Environmental Regulations Work in Kenya?: A Case Study of the Thika Highway Improvement Project. Center for Sustainable Urban Development.

Calignano, G. and Vaaland, T.I., 2018. Local Content in Tanzania: Are Local Suppliers Motivated to Improve? *The Extractive Industries and Society*, 5(1), pp. 104–113.

Cappelen, Å. and Mjøset, L., 2009. Can Norway Be a Role Model for Natural Resource-Abundant Countries? Chap., 3, pp. 44–72.

Cavalheiro, G.M. do C., Joia, L.A., & Gonzalves, A.C., 2014. Strategic Patenting in the Upstream Oil and Gas Industry: Assessing the Impact of the Pre-salt Discovery on Patent Applications in Brazil. *World Patent Information*, 39, 58–68.

Chrispine Odour et al., 2014. The Political Economy of the Extractive Sector. In Miriam, W., Omolo, O., and Mwabu, G., eds, *A Primer to the Emerging Extractive Sector in Kenya: Resource Bliss, Dilemma or Curse.* Nairobi: Institute of Economic Affairs, p. 181.

Cruz, A.S. and Mafambissa, F.J., 2016. Industries Without Smokestacks. Industries without Smokestacks, p. 232.

Engen, O.A., 2009. The Development of the Norwegian Petroleum Innovation System: A Historical Overview. *Innovation, Path Dependency and Policy: The Norwegian Case*, pp. 179–207.

Filho, N.N. Brazil's Oil & Gas Local Content Policy: Lessons Learned. Institute of the Americas.

Fraser, N., 2009. Social Justice in the Age of Identity Politics. Geographic Thought: A Praxis Perspective, pp. 72–91.

Hansen, M.W., 2013. Reaping the Rewards of Foreign Direct Investment: Linkages between Extractive Multinational Corporations and Local Firms in Tanzania.

Hansen, M.W., Buur, L., Mette Kjær, A. and Therkildsen, O., 2016, May. The Economics and Politics of Local Content in African Extractives: Lessons from Tanzania, Uganda and Mozambique. In the *Forum for Development Studies*, 43(2), pp. 201–228. Routledge.

Heffron, R.J. and McCauley, D., 2017. The Concept of Energy Justice Across the Disciplines. *Energy Policy*, 105, pp. 658–667.

Heum, P., 2008. Local Content Development: Experiences from Oil and Gas Activities in Norway.

Ihua, U.B., Ajayi, C. and Eloji, K.N., 2009, May. Nigerian Content Policy in the Oil and Gas Industry: Implications for Small and Medium-sized Oil Services Companies. In Proceedings of the 10th Annual Conference, IAABD.

Jenkins, K., McCauley, D., Heffron, R., Stephan, H. and Rehner, R., 2016. Energy Justice: A Conceptual Review. *Energy Research & Social Science*, 11, pp. 174–182.

Kjær, A.M., and Katusiimeh, M., 2012. Growing But Not Transforming: Fragmented Ruling Coalitions and Economic Developments in Uganda (No. 2012: 07). DIIS working paper.

Klueh, U.H., Pastor, G., & Segura, A., 2009. Policies to Improve the Local Impact from Hydrocarbon Extraction: Observations on West Africa and Possible Lessons for Central Asia. *Energy Policy*, 37(3), pp. 1128–1144.

Lee, B. and Dupuy, K., 2018. Understanding the Lie of the Land: An Institutional Analysis of Petro-Governance in Tanzania. *Journal of Energy & Natural Resources Law*, 36(1), pp. 85–101.

Lima de Oliveira, R., 2016. Resource-Led Industrial Development in the Oil & Gas Supply Chain: The Case of Brazil.

Local Content Bill of Kenya, 2018.

Local Content Bill, 2017.

Local Content: Goals and Brief History. Rio de Janeiro: ANP. http://www.anp.gov.br/brasil-rounds/round10/ingles/conteudo_local_Obj_e.asp.

Maxon, R., 2009. *East Africa: An Introductory History*. West Virginia University Press.

Mendonça, R.W. and de Oliveira, L.G. 2013. Local Content Policy in the Brazilian Oil and Gas Sectoral System of Innovation. *Latin American Business Review*, 14(3–4), pp. 271–287.

Mining (Local Content)(Amendments) Regulations, 2019.

Mkutu, K., Mkutu, T., Marani, M. and Ekitela, A.L. 2019. New Oil Developments in a Remote Area: Environmental Justice and Participation in Turkana, Kenya. *The Journal of Environment & Development*.

Mohammed, I. 1985. The Nigerian Enterprises Promotion Decrees (1972 and 1977) and Indigenisation in Nigeria (Doctoral dissertation, University of Warwick).

Murungi, C.N. Introducing Local Content Regulation in Kenya's Economy: A Case Study of Kenya's Energy, Oil and Gas Sector. Institute of Diplomacy and International Studies. The University of Nairobi. R50/63843/2010

Nigerian Oil and Gas Industry Content Development Act, 2010.

Noreng, Ø. 2009. Brazil and Norway–Offshore Petroleum Experiences and Lessons. *The Journal of Energy and Development*, 35(1/2), pp. 79–99.

Nwaobi, G.C. 2004. Corruption and Bribery in the Nigerian Economy. Available at SSRN 506342.

Nwapi, C. 2015. Defining the "Local" in Local Content Requirements in the Oil and Gas and Mining Sectors in Developing Countries. *Law and Development Review*, 8(1), pp. 187–216.

Obiri, K.A., Bjeirmi, B., Boateng, P. 2019. Local Content Implementation Enhancement Through Infrastructure Development in Ghana's Oil and Gas Industry. *Journal of Energy Research and Reviews*, pp. 1–10.

Ogbuagu, C.S. 1983. The Nigerian Indigenization Policy: Nationalism or Pragmatism? *African Affairs*, pp. 241–266.

Okoli, A.C., 2019. Oil Pipeline Vandalism in the Niger Delta: Need, Greed and Grievance Factors.

Ovadia, J.S., 2016. Local Content Policies and Petro-Development in Sub-Saharan Africa: A Comparative Analysis. *Resources Policy*, 49, pp. 20–30.

Ovadia, J.S. 2017. Local Content in Tanzania's Gas and Minerals Sectors: Who Regulates? CMI Brief.

An Overview of the Brazilian Oil and Gas Industry. http://www.offshorecenter.dk/filer/files/Project/Internationalisering/OCD%20report%20Brazil.pdf.

Oyewole, B. 2018. Overview of Local Content Regulatory Frameworks in Selected ECCAS Countries.

Paul, I.A. 2015. A Historical Perspective of Petroleum on Nigeria's Economy Crisis Since Independence. *Global Journal of Human-Social Science Research*, 15, pp. 17–24.

Raji, A.O.Y. and Abejide, T.S., 2014. The British Mining and Oil Regulations in Colonial Nigeria c. 1914–1960s: An Assessment. *Singaporean Journal of Business, Economics and Management Studies*, 51(1447), pp. 1–14.

Ramírez-Cendrero, J.M. and Wirth, E., 2016. Is the Norwegian Model Exportable to Combat Dutch Disease? *Resources Policy*, 48, pp. 85–96.

Schlosberg, D., and Carruthers, D. 2010. Local Struggles, Environmental Justice, and Community Capabilities. *Environmental Politics*, 10(4),pp. 12–35.

Sen, R. 2018. Enhancing Local Content in Uganda's Oil and Gas Industry (No. 2018/110). WIDER Working Paper.

Steinar Holden, Avoiding the Resource Curse the Case Norway Department of Economics (University of Oslo, June 2011).

The Petroleum (Exploration, Development and Production) Act of Uganda, 2013.

Tina Hunter, Law and Policy Frameworks for Local Content in the Development of Petroleum Resources: Norwegian and Australian Perspectives on Cross-Sectoral Linkages and Economic Diversification, 27 MINERAL ECONOMICS 115 (2014).

Tina Hunter, Legal Regulatory Framework for the Sustainable Extraction of Australian Offshore Petroleum Resources. A Critical Functional Analysis (The University of Bergen, 2010).

Tordo, S., Warner, M., Manzano, O. and Anouti, Y. 2013. *Local Content Policies in the Oil and Gas Sector*. The World Bank.

The United Republic of Tanzania, Local Content Policy (2014).

6

Creating a Regional Content Policy

6.1 Introduction

Sub-Saharan Africa has a challenge of cracking through the enclave and transforming their oil and gas resources into a vehicle of sustainable development for individual countries in the region. The ramifications of adopting and implementing a national content policy in developing Sub-Saharan countries have been seen in Nigeria. The national content of Nigeria and Angola have built a corrupt cluster within the political elite, coupled with low skilled labour without a pre-existing industry were skills and expertise could be leveraged to develop the oil and gas sector. The national content policy had become redundant in its application. This called for numerous reforms since the indigenisation policy of the 1970s up to the National Content Act of 2010. Sub-Saharan African countries do not have the features in their politics, infrastructure as well as in their economies to replicate the Norwegian or Brazilian model of national content success. This calls for the need for collaboration between Sub-Saharan oil and gas economies, such as those in the Gulf of Guinea, the East African Community, the Southern African Development Cooperation and the Common Market for Eastern and

R. S. Muhongo, *Energy Justice*, Energy, Climate and the Environment,
https://doi.org/10.1007/978-3-030-61338-9_6

Southern Africa. The case study shows that the costs of domestic self-reliance in cracking through the enclave and integrating the oil and gas value chain are excessive.[1] The enclave nature of the oil and gas industry calls for collaboration between Sub-Saharan Africa to meet the demands of the oil and gas industry. This stems from the belief we are all citizens of the world, and so have we considered the effects beyond our borders and from a global context.[2] Sub-Saharan economies are vulnerable due to the inability to govern and integrate the oil and gas value chain.[3] These countries have generally smaller consumer markets with less depth (both in terms of the quantity and range of local industries) which can be used as tools to crack the enclave and integrate the oil and gas value chain. This gives room for the big multinational oil and gas companies to abuse the market.

Most Sub-Saharan African countries may not have the absorptive capacity or may not have the technology and expertise in dealing with the multinational oil and gas companies. Hence developing countries may be an easy target for international cartels because of their weaker enforcement mechanisms towards local content policies because developing countries cannot individually have the capacity to integrate the oil and gas value chain as well as channel the large extractive FDI into developmental projects that can be advantageous for the whole country at large. Countries in regional developmental blocs such as Kenya, Uganda and Tanzania; Angola, Ghana, Equatorial Guinea, Liberia and Nigeria must rely on cooperative institutions and mechanisms that could leverage their competitive and comparative advantages to integrate the oil and gas value chain. This chapter discusses the creation of a regional content policy as a typology of local content policy, and the chapter provides the analytical and contextual framework of regional content policy.

---

[1] Haas, E.B., 1980. Why Collaborate? Issue-Linkage and International Regimes. *World Politics*, 32(3), pp. 357–405.

[2] Heffron, R.J., 2020. The Role of Justice in Developing Critical Minerals. The Extractive Industries and Society.

[3] Ngoasong, M.Z., 2014. How International Oil and Gas Companies Respond to Local Content Policies in Petroleum-Producing Developing Countries: A Narrative Enquiry. *Energy Policy*, 73, pp. 471–479.

6.2 Defining the Term "Regional Content"

> Regional content: entails the government intervention through regional integration tools such as cooperative and competitive mechanisms to create a deliberate first preference and participation in the provisions for goods and services in the extractive industry.[4]

The basic idea behind the regional content policy is the aggregation of capital, knowledge, technology, goods and services among Sub-Saharan African resource-rich countries. The regional content policy calls for different degree of intervention to take place at the national level and even regional level due to the policy. There must be a free movement of goods and services. As well as the right of establishment, the East African Community and Economic Community of West African States has to take over with regards to local content policy harmonisation other interventions in "national" markets should be distortive. Regional content development must consider the quality of relationships between the political, economic, social, infrastructural, environmental, and local capability aspects of an economy and their relation to the industrial, procurement, and public or national policies.[5] The level of regional intervention must foster policy and inter-sectoral coordination aligned to the region's specific characteristics to implement provisions for the preference of goods and services.

The regional content approach concentrates on value addition. Warner defines local value added as "the wealth local companies create in transforming materials and services purchased from other countries into revenue-generating output".[6] The value addition must encompass a wider approach due to the multiple actors and stakeholders involved. In a regional content framework, value addition must entail enhancing

---

[4]Developed by author 2019.

[5]Ramdoo, I., 2015. Synergising and Optimising Mineral Infrastructure in Regional Development Strategies. By International Centre for Trade and Sustainable Development (ICTSD) 7.

[6]Warner, M., 2011. *Local Content Solutions: Participation of Domestic Industry in Procurement for Oil, Gas and Mining Projects*. Sheffield: Greenleaf Publishing.

productivity of other economic sectors, supporting diversification in the region as well as improving regional and national competitiveness.

This competitive advantage gained in a region refers to where the optimal structural relations between production factors in changing conditions are used to improve the living standards and value chain integration of local stakeholders in their national oil and gas value chains.[7] Furthermore, local participation in a regional content framework entails different countries agreeing among themselves to concentrate on a few strategic industries in which to set higher targets so that their local firms might be granted first preference across the region because they are competitive in price, quality and have timeous delivery.

It has already been established that for there to be local content; there must be governmental intervention. In the case of a regional content policy there is a regional intervention, this entails the integration of regional actors (through partnerships, Agreements, clusters and/ cooperatives) in the value chain of a national oil and gas industry. Sustainable cooperation in a regional content policy entails aligning policies and practices of the different countries within a region to have a unified vision and developmental plan.

As a region, the policy facilitates the region to gain a competitive advantage over other regions (in this case, regions like ECOWAS).[8] The East and West African countries will obtain a better environment that can foster skills and better supplier development schemes that will be more efficient and effective in comparison to other regions in Africa. The result is attracting more FDI in the region due to the regional policy. The comparative advantage to produce goods and services at a lower

[7]Budd, L. and Hirmis, A., 2004. Conceptual Framework for Regional Competitiveness. *Regional Studies*, 38(9), pp. 1015–1028.

[8]Competitive advantage may be defined as: the ability to implement a competitive strategy so as to secure a favourable position. Clemons, E.K., 2019. Resources, Platforms, and Sustainable Competitive Advantage: How to Win and Keep on Winning. In *New Patterns of Power and Profit* (pp. 93–104). Cham: Palgrave Macmillan.

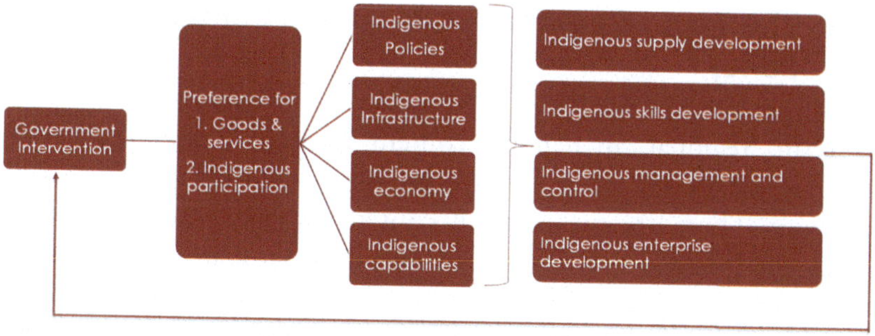

Fig. 6.1 Schematic representation of government intervention in a regional content policy (*Source* Author's elaboration)

cost while maintaining business development. With all these developments, the policy offers the governing and assuring structure of the policy between the adopting countries.

The schematic representation in Fig. 6.1 I shows the channels of regional content policy through governmental intervention. For oil and gas resource-rich countries, the aim is to have local first preference for goods and services. The regional content policy functions through the harmonisation of local policies such as depletion policy, industrial policy and the procurement policy as well as policies that affect overall extractive sector development. The local, regional economy must create the absorptive capacity that can create the local preference based on their pooled resources. The local capabilities and infrastructure in the region must be used collectively to facilitate the regional intervention mechanisms in the oil and gas value chain. In East Africa the East African Crude Oil Pipeline is an example of infrastructure being developed by East African countries, the pipeline has created numerous benefits for the local communities within East Africa (specifically Tanzania and Uganda). The regional content policy has only four aims, and objectives and these are to develop local supply development, local skills, local management and control of oil and gas resources as well as local enterprise development.

6.2.1 The Term "Regional" in Regional Content Policy

The term "regional" in regional content must be determined for the proper regulation of the policy. Local content can be applied in three typologies. These are the national content, whereby the term "local" would mean all the local people coming from any area of the nation. The community content, the term "local" would mean the local people that come from the neighbouring areas of the oil and gas operations. The regional content policy, the term "regional" would mean any person originating from the countries that are member states of the regional economic bloc implementing a regional content framework. The term "local" has led to some confusion with regards to local content development in resource-rich countries. In East African countries were people are strongly ethnocentric; people are more likely to identify themselves according to their ethnicity rather than their nationality.[9] Uganda presents the most extreme form of ethnicity identification in the region, having the highest index of ethnic fractionalisation in the world.[10] In Tanzania, there is also significant additional tension between black Africans (who dominate political power) and Asian and Arab Tanzanians (who dominate the business sector).[11] In Kenya, their tension also raises between the Kikuyu (the ruling tribe of Kenya) and the Luo (the major opposition tribe in Kenya). Ironically the resources are found in the locality of the poorest and most marginalised community in Kenya known as the Turkana. It is evident that due to the fractionalisation of the different ethnicities in East Africa. The adopted national content approach will not reach its full potential due to fractionalisation. The

[9] Nwapi, C., 2015. Defining the "Local" in Local Content Requirements in the Oil and Gas and Mining Sectors in Developing Countries. *Law and Development Review*, 8(1), pp. 187–216.

[10] Alesina, A., et al., 2002. Fractionalization. Harvard Institute of Economic Research Discussion Paper No 1959, p. 8, online: http://dash.harvard.edu/bitstream/handle/1/4553003/alesinassrn_fractionalization.pdf?sequence=2.

[11] Booth D., et al., East African Prospects: An Update on the Political Economy of Kenya, Rwanda, Tanzania and Uganda. ODI Report, May 2014, p. 35, online: http://www.odi.org/sites/odi.org.uk/files/odi-assets/publications-opinion-files/8945.pdf.

policy will be defined along ethnicity lines, and this will result in the creation of a local capitalist state.[12]

We define the term "regional" as any person originating from countries that are member states of the regional economic bloc. Implementing a regional content policy offers social stability in the region that automatically leads to national and regional unity with regards to attaining first preference in the oil and gas value chain. Ideally, the regional framework is designed to create more winners as it will bring about concurrent economic development. Furthermore, the term "regional" will also be applicable when it comes to companies or firms originating within a given region. The normal criteria of assessing whether a firm originates from a certain locality such as means through local registration, local ownership, local workforce, local value-added and joint venture.[13] These criteria will not be applicable in determining what is regarded as "regional" when defining a "regional firm". This is because a regional content policy ought to avail a level playing field for both the local firms and the foreign investors through different mechanisms such as creating clusters, special economic zones, industrial parks, and resource corridors etc.

> Hence a regional company should be regarded as a company that operates within the given region, and 51% of the equity share of the company must be owned by natural citizens originating from the given region.

The agglomeration of different entities to create a regional entity in a regional content policy creates a comparative advantage, and the economics of competition comes into play. John Jackson state that "the economists themselves have articulated certain challenges to the concepts of comparative advantage".[14] Even amidst the challenges the advantages

[12]The creation of a local capitalist state has been seen in countries such as Angola and Nigeria were the ruling ethnic group controls the resources and uses the national content policy to finance patrimonial acts and clientelism in these respective jurisdiction. Ovadia, J.S., 2013. The Making of Oil-Backed Local Capitalism in Nigeria. *New Political Economy*, 18(2), pp. 258–283.

[13]Estevez, A. M., et al., 2013. Local Content Initiatives: Enhancing the Subnational Benefits of the Oil, Gas and Mining Sectors (RWI).

[14]Jackson, J.H., 2000. The Perils of Globalization and the World Trading System. *Fordham Int'l LJ*, 24, p. 371.

of having a comparative advantage such as attaining a higher degree of efficiency and productivity that would contribute to the betterment of regional economics due to the competition between different producers and service providers within the regional content policy.

6.3 The Rationale Behind Adopting a Regional Content Policy

6.3.1 The Decreasing Importance of National Policies

The constraints in developing national and community content policies in Kenya, Tanzania and Uganda, bring about the incentive towards increased attention to seek solutions from the regional economy. These East African countries have rendered part of their national sovereignty on socio-economic and political policy to international agreements. These are the East African Community (EAC), the African Continental Free Trade Area (Acfta), the Common Market Eastern and Southern Africa Community (COMESA) and the Southern Africa Democratic Cooperation (SADC). The oil and gas industry is an industry with global value chains that predict the value of the commodities internationally.[15] It is difficult for local content policies adopted by poor weak economies with a fragmented national policy framework, to use the conventional fiscal and monetary measures availed to central governments for policy efficiency.[16] Because local content policies are expensive to implement, it becomes increasingly difficult for these national economies to support particular industries or firms.[17] Countries with a strict budget and over-reliance on donor dependence such as Tanzania, Kenya and Tanzania are reluctant to individually commit themselves to expensive strategic policies that will automatically lead to major public expenditure. A Regional

[15]Emmanuel, R., 2016. Optimal Local Content Requirement Policies for Extractive Industries. *Resources Policy*, 50, pp. 244–252.

[16]Ibid.

[17]Ovadia, J.S., 2016. Local Content Policies and Petro-Development in Sub-Saharan Africa: A Comparative Analysis. *Resources Policy*, 49, pp. 20–30.

content policy approach can facilitate local governments increase activities in promoting investments and enterprise competitiveness in the local (indigenous) areas of the oil and gas operations as well as the whole nation. The regional level can also facilitate the mobilisation of policy tools to adopt an adequate regional content framework for East African countries.

6.3.2 Market Failure

In poor weak economies with oil and gas resources, market failure becomes a fundamental rationale for adopting a regional content framework. Weak economies like Kenya, Uganda and Tanzania have market imperfections due to asymmetric information and economies of scale, among many other reasons. Local content policies in East Africa do not meet the market imperfections of the oil and gas industry. Inadequate R&D measures undertaken by both the government and IOCs under the local content umbrella do not function in these economies.[18] This is because the social rate of return is expected to be higher than the private gain for the IOCs in developing R&D initiatives. Thus adopting a regional content framework facilitates knowledge creation, creates appropriate channels for information transfer with higher levels of accountability. The regional policy can act as a risk management initiative while implementing regional content policy requirements. A regional content policy can also be directly or indirectly motivated by the inability of the East African economies to adopt and sustain inter-linkages within the oil and gas value chain that are potentially favourable for the whole society. For market failure to serve as a rationale for regional intervention, it is necessary to have a clear understanding of why markets are unable to bring about desirable outcomes.

[18]Lange, S. and Kinyondo, A., 2016. Resource Nationalism and Local Content in Tanzania: Experiences from Mining and Consequences for the Petroleum Sector. *The Extractive Industries and Society*, 3(4), pp. 1095–1104.

Furthermore, market failure also prevails due to a lack of coordination as well as associated externalities.[19] Local content policies are highly susceptible to corruption. Most integrated firms in the oil and gas value chain are normally politically affiliated to patronage in one way or the other, especially in African economies.[20] This brings about a "prisoners' dilemma" whereby these firms are reluctant to cooperate with other local firms despite the probable gains.

The patrimonial and corrupt nature of national and community content policies hinders investments in capacity building, brings about minimal initiatives in creating networks as well as attracting external players from global markets. A regional content framework has a transparency that creates the legitimacy to integrate global value chains as the inefficiencies of one country do not constrain it. This will result in higher spill-over effects within the region that will attract more investments.

6.3.3 Systematic Failure in Sub-Saharan African Oil and Gas Industry

Systemic failure takes place when there is any inconsistency between these interconnected institutions, organisations.[21] Local content implementation and development is leveraged on existing capabilities enhanced by the interconnected institutions and other regulatory bodies. These capabilities can be within manufacturing, fabrication, and services. However, Sub-Saharan African government's intervention hinders the proper functioning of market forces. In Kenya, Ghana, Nigeria, Angola, Tanzania and Uganda, market-based inputs replace public inputs.[22] There are functions that markets cannot perform, such as establishing

---

[19] Farooki, M. and Kaplinsky, R., 2014. Promoting Diversification in Resource-Rich Economies. *Mineral Economics*, 27(2–3), pp. 103–113.

[20] Lange, S. and Kinyondo, A., 2016. Resource Nationalism and Local Content in Tanzania: Experiences from Mining and Consequences for the Petroleum Sector. *The Extractive Industries and Society*, 3(4), pp. 1095–1104.

[21] Abidin, M.Z., 2001. Competitive Industrialization with Natural Resource Abundance: Malaysia. In R.M Auty (Ed.), *Resource Abundance and Economic Development* (pp. 147–164). Oxford University Press.

[22] Jacob, T. and Pedersen, R.H., 2018. New Resource Nationalism? Continuity and Change in Tanzania's Extractive Industries. *The Extractive Industries and Society*, 5(2), pp. 287–292.

company registries, setting norms, enforcing contracts and regulatory frameworks as well as providing infrastructure.[23]

These country's monopolies that dictate how the public inputs should be implemented to their favour in this is detrimental to the implementation of policies such as community and national content policies. This leads to systematic failure because different activities call for different kinds of intervention, and governments fail to predict when intervention is justifiable or not, especially in the oil and gas industry. Adopting a regional content framework ensures interactions between different actors as well as different driving forces.

6.4 Characteristics Needed for the Success of a Regional Content Policy

A regional content policy must adopt the characteristics of a single market, though the policy in itself is adopted and implemented within a common market. The general characteristics of this policy are such as the free movement of goods and services, the aggregation of skills and technical abilities, the proliferation of clusters and industrial parks, the creation of transport corridors as well as the linkage of different tertiary and vocational training institutions within the region.[24] The following characteristics are important for the success of the regional content policy.

6.4.1 The Regional Content Policy Must Be Adopted in a Rule-Oriented System

The regional content policy's cooperative mechanism must have a rule-oriented system with treaty rules that are justifiably effective. The presence of a rule-oriented system is important because a regional policy

[23]Humphreys, M., Sachs, J.D and Stiglitz, J.E. (Eds.), 2007. *Escaping the Resource Curse*. New York: Columbia University Press.

[24]Bulmer, S.J., 1998. New Institutionalism and the Governance of the Single European Market. *Journal of European Public Policy*, 5(3), pp. 365–386.

needs a foundation from which it can draw governmental actions that can avail remedies to market failure. A regional content policy cannot have the power to implement fiscal measures, nor subsidise the market. The regional content policy rearranges the market structures, but the policy can only function through governmental intervention.

The role of the policy is regulatory through the use of rules to implement measures that create a deliberate first preference for the local population within the region. The mere establishment and creation of these rules are not enough. The rules must be implementable and effective as well as justifiable.

6.4.2 Dispute Settlement Procedures

The effective rule-based system must be effective, and the rules must be implementable, especially in dispute settlement procedures. This is because a regional content policy does not have the same type of monopoly of power that the national content often has and thus cannot enforce the rules using any force or sanctions. The regional content policy will be dependent on the persuasive effect of procedures and diplomacy. In a national and community content policy regulations of the policy are easily obeyed since the multinational oil companies want to be treated fairly within the ambit of the national laws. But for regional policy, the techniques of enforcement to be successful, the regional framework and its regulations must be creditable and deemed legitimate.

6.4.3 Credibility and Legitimacy

Significantly, the procedures of the regional policy will lead to the enhancement of the critically needed characteristics of credibility and legitimacy. These characteristics that are called for by the significance of the variety of stakeholders include transparency and participation.

6.4.4 Transparency of the Regional Policy

The current technological advancements have heightened both the desire and the need for greater access to information for stakeholders. The different stakeholders of the regional content policy will have the desire to have enough knowledge on the pertinent issues such as the tender procedures, skills advance techniques and other opportunities availed to the local to integrate the oil and gas value chain. Stakeholders will also wish to know the reasoning of decisions and what the influences on those decisions were. This is a critical "check and balance" on governments' misuse of power, and any ill practices done by the multinational oil companies.

6.5 Regional Content Policy in Comparison to Other Instruments

Implementing a regional content policy can be misconstrued as any other instrument that is employed in regional economic blocs due to regional integration. This can be because most local content regimes have implemented national and community content policies. Hence it is important to create clarity on what a regional content policy entails apart from similar policies such as industrial policies, regional value chains and special economic zones.

6.5.1 Regional Content and Regional Industrial Policy

A regional content approach differs fundamentally from industrial policies adopted by regional bodies in ways that they are founded and the implications of both policies. Industrial policy is based on the notion of protecting local firms from international competition.[25] This is because

[25]Jesus, F.D., Silva, J.F. and Roitman, T., 2017. Industrial Policy and Local Content: Brazil's Pre-Salt Area Case. Offshore Technology Conference.

some industries offer better prospects than others within a country. Hence they must be protected until they acquire the necessary scale of sophistication to function on their own.

A regional content policy, in contrast, applies in a more vigorous manner of competition among firms within a region and their coordination thereof. A regional content policy does not implement entry barriers (against and local or foreign firms). Instead, a regional policy stresses the need for member countries to cooperate and create a competitive advantage that will not be affected by the steady opening of the local market to imports. This results in improvement of supply chain efficiency, upgrades the local capabilities and stimulates a healthy rivalry within the oil and gas value chain across the region.

6.5.2 Regional Content and Special Economic Zones

SEZs these are focused geographic companies. They are established to bring about infrastructural development and enhance R&D.[26] SEZs do not give government incentives that are found outside the geographic concentrated zones. SEZs are traditionally created by explicit industrial policy intervention to promote national economic growth.[27] These SEZs can be high tech parks, science parks, industrial zones and export processing zones are examples. A regional content policy is less a top-down approach, and it is not limited to physical or hard infrastructure. Supportive institutions and R&D facilities evolve gradually to respond adequately to the international market so that the local firms within the region can meet the sophisticated demands of the industry. The role of the government in implementing a regional content framework is to avail the means of integration and cooperation in a conducive business environment. This policy is not limited to a particular sector, but as a country in general to further linkages and growth of supporting sectors to the oil and gas industry. Another point to be mentioned is the fact that SEZs are limited to given geographic concertation while a regional content policy

[26]Gereffi, G., Humphrey, J. and Sturgeon, T., 2005. The Governance of Global Value Chains. *Review of International Political Economy*, 12(1), pp. 78–104.

[27]Ibid.

can be spread over multiple nations or regions of different geographic locations dependent on the agreement unifying them.

6.5.3 Regional Content and Regional Value Chains

A value chain approach is a linear transaction-oriented approach that focuses on transactional efficiency within a chain.[28] A regional content approach is more systematic, and it entails analysis of the whole value chain for local integration. Furthermore, the policy develops strategies for adequate spatial orientation and focuses on solving coordination and information failures.[29] This is enabled through better participation of supporting institutions outside the value chain. Value chain projects are directed towards helping specific beneficiaries. Such as the marginalised groups sunk in poverty.[30] The regional content policy is integrated through the entire value chain. This would include all segments of the oil and gas industry from authorisation all the way to the decommissioning or waste management phase of the project.

6.6 The Guiding Principles of a Regional Content Policy

The guiding principles of a regional content policy will have its foundation entrenched in energy justice. The reason for adopting a regional content policy within the framework of energy justice is because consideration needs to be taken when adopting local content policies. The consideration undertaken ensures that policies, plans and programmes that are implemented under the new regional content policy guarantee fair and equitable access to resources and opportunities that are availed

[28]Gereffi, G., Humphrey, J. and Sturgeon, T., 2005. The Governance of Global Value Chains. *Review of International Political Economy*, 12(1), pp. 78–104.
[29]Ibid.
[30]Gereffi, G., Humphrey, J. and Sturgeon, T., 2005. The Governance of Global Value Chains. *Review of International Political Economy*, 12(1), pp. 78–104.

due to the oil and gas industry.[31] The principles of energy justice ensure that local content is a policy that does originate from nascent oil and gas countries like Tanzania, Kenya and Uganda. But will account for distributional, procedural and recognition inequalities as well as an emerging non-Western understanding of the implementation of local content policies as well as the development of oil and gas industry.[32] The eight principles of energy justice will be applicable in a regional content policy in the following manner:

6.6.1 Availability

The people that are directly and indirectly affected by the oil gas operations will be given equal opportunity to integrate the oil and gas value chain through a deliberate preference that is availed by the common market under which the regional policy is attached to. Hence if the East African countries adopt and implement a regional content policy, countries such as Kenya, Uganda and Tanzania all companies and peoples of these countries will be given equal opportunity as the principle of "availability" envisages.

6.6.2 Affordability

The regional content policy will facilitate the delivery of goods and services at affordable and competitive rates. Whilst delivering goods and services the regional policy will also ensure that the tendering procedures that have been adopted are affordable and easily accessible to all members of the regional policy.

---

31 McCauley, D., Ramasar, V., Heffron, R.J., Sovacool, B.K., Mebratu, D. and Mundaca, L., 2019. Energy Justice in the Transition to Low Carbon Energy Systems: Exploring Key Themes in Interdisciplinary Research.

32 McCauley, D., Ramasar, V., Heffron, R.J., Sovacool, B.K., Mebratu, D. and Mundaca, L., 2019. Energy Justice in the Transition to Low Carbon Energy Systems: Exploring Key Themes in Interdisciplinary Research.

6.6.3 Due Process

The countries and parties to the regional content policy must follow a due process and respect human rights as well as make sure not to violate any laws, rules and regulations in that regard.

6.6.4 Transparency and Accountability

The countries and parties involved in a regional content policy must have all the relevant information about the oil and gas industry and the opportunities availed to the members of the regional content policy. The policy must be fair and must have mechanisms of holing one accountable in cases of any misdemeanour. The policy must also inform its members and parties on all decision-making process within the framework.

6.6.5 Sustainability

The regional content policy must not only concentrate on oil and gas ventures, skill-building facilities and technological advancement. The policy must avail opportunities in other industries through linkages and other created corridors to develop the affected communities as well as maintain the environment.

6.6.6 Intra-Generational Equity

The regional content policy must award fair and equal access to all members and parties that are part of the regional content policy. In National and community content policies the remote regions that usually have the oil and gas resources within their vicinity are fundamentally marginalised communities that do not have access to the opportunities of the national and community content policy. This has developed into "Not in My Back Yard syndrome" that has affected numerous energy operations in different regions. The regional content policy covers all people that have rights to the policy since in a regional level their numerous remote areas to be dealt with and not just one or two oil and gas-rich remote communities.

6.6.7 Inter-Generational Equity

The regional policy must implement the goals of the policy namely technological transfer from multinational oil companies to local firms and institutions, local enterprise development, control and monitoring of oil and gas resources, as well as a local intervention into the oil and gas value chain. These goals must be met while taking account of future generations. Hence industries in renewable energy must also be developed to as to ensure that future generations rip the fruits of the regional content policy.

6.6.8 Responsibility

The parties and members of the regional content policy must act responsibly while taking into account the wellness of the environment, preserving nature as well as making sure the places of oil and gas operations can be rehabilitated after the operations are completed.

6.7 Advantages and Disadvantages of a Regional Content Policy

Local content policies are known for serving a dual-function of being both advantageous and disadvantageous. Though these ramifications are due to other numerous factors that are not only limited to the oil and gas industry but to the whole governance structure of the country in question. Since regional content is a typology of local content policies though the policy has not yet been implementable in any oil and gas region, the case study analyses the probable disadvantages and advantages of the regional content policy.

6.7.1 Regional Linkages

A regional content framework creates wider opportunities that facilitate deeper linkage formation through regional cooperation. The larger the pool of talent in a region, the more foreign investors will be attracted to the region.[33] When operating larger economies of scale in a nation, the local suppliers in poor economies are usually not able to meet the demands of the oil and gas industry. Hence a regional procurement scheme adopted by a regional content policy could facilitate the standardisation of what local content entails within the region. Capital intensive and high capacity downstream projects, such as smelters, are more likely to be viable if viewed from a regional rather than national perspective.[34] The adoption of regional training institutes can also minimise the costs for the national economies. Infrastructural needs are easier to meet at a regional level rather than a national level.

6.7.2 Address Institutional Failures

Many resource-rich countries are characterised by institutional failures, constraining local content development in a nation.[35] Due to the existence of a myriad of government agencies regulating local content frameworks in resource-rich countries. There is often a lack of coordination between the relevant ministries and government agencies and IOCs as well as among the different governmental bodies.[36] A regional content policy avails a coherent and coordinated policy approach in the region that is crucial for the development of the oil and gas industry. This can be facilitated by the creation of a strong independent regional regulatory authority that makes impartial decisions with regards to regional content

[33]Morrissey, O., 2012. FDI in Sub-Saharan Africa: Few Linkages, Fewer Spill-Overs. *European Journal of Development Research*, 24(1), pp. 26–31.

[34]Hansen, M.W., 2014. From Enclave to Linkage Economies? A Review of the Literature on Linkages Between Extractive Multinational Corporations and Local Industry in Africa, DIIS Working Paper, no. 2, Copenhagen, Danish Insitute for International Studies.

[35]Ibid.

[36]Ovadia, J.S., 2017. Local Content in Tanzania's Gas and Minerals Sectors: Who Regulates? CMI Brief.

development. Due to the expensive nature of upstream operations, a dedicated regional body can devise mechanisms through pooled skills, capital and facilitate the implementation of different local and regional content requirements in the upstream sector throughout the region.[37]

6.7.3 Minimises Corruption and Patronage in the Oil and Gas Industry

Resource-rich countries usually have an unfavourable political economy that is characterised by clientelism and patronage.[38] These countries like Nigeria, Angola and Equatorial Guinea lack principles of good governance and transparency in their oil and gas industry. Large cash flows and rent-seeking tendencies characterise the oil and gas industry.[39] The development of a local/national content policy is a long-term strategy that results cannot be readily availed within one election cycle. These elements of the oil and gas industry bring about incentives for the political elite to create mechanisms for personal benefit from the short-term oil and gas agreements or infrastructure solutions that are sub-optimal form an economic and regional perspective.[40] Such corrupt practices impede long-term strategies implemented for local content development. Adopting a regional content approach offers a longer term broad-based approach that is not limited to ill practices of one country. Thus, development partners in a regional content policy can implement the requirements of a regional content policy in a way that reflects good practices while taking into account the local contexts, institutions and politics.

In countries where corruption is high, and governance is weak it is unlikely to have the "political will" that will create a sustainable local

---

[37] International Finance Corporation (IFC), 2015. The Art and Science of Benefit Sharing in the Natural Resource Sector, A discussion paper. International Finance Corporation. Washington, DC.

[38] Ovadia, J.S., 2013. The Making of Oil-Backed Local Capitalism in Nigeria. *New Political Economy*, 18(2), pp. 258–283.

[39] Ibid.

[40] Ovadia, J.S., 2014. Local Content and Natural Resource Governance: The Cases of Angola and Nigeria. *The Extractive Industries and Society*, 1(2), pp. 137–146.

content policy, which will not be manipulated by the selected few in the political elite.[41] In such cases, a regional content policy may be adopted and implemented to place greater emphasis on the development institutions and the promotion of transparency in the oil and gas sector.

6.8 Disadvantages of a Regional Content Policy

6.8.1 Strained Relations Within the Region

Regional content is a noble cause, but the policy may lead to disaffection to certain stakeholders in implementing the policy. The local business magnets may not be pleased with adopting a regional content policy. The fear competition from neighbouring countries within the region may lead to strained relations. In developing countries, the local population might fear the sudden migration into these remote areas.[42] This may result in strained relationships between the local and the neighbouring market participants, such as bringing about xenophobic attacks.

6.8.2 Increased Barriers

A regional content policy might increase trade barriers against members who are not included in the regional content policy. When this happens, the countries that are not part of the regional content policy will divert their trade to other countries. This can lead to higher costs of production. East Africa, as an example, the economic bloc, the East African Community (EAC) is made up of six countries, namely Tanzania, Uganda, South Sudan, Kenya, Rwanda and Burundi. The countries suggested to adopt a regional content framework are Tanzania, Kenya and Uganda because

[41] Ibid.

[42] Johannes, E.M., Zulu, L.C. and Kalipeni, E., 2015. Oil Discovery in Turkana County, Kenya: A Source of Conflict or Development? *African Geographical Review*, 34(2), pp. 142–164.

they have similar natural resources and circumstances.[43] Other countries like Rwanda and Burundi that don't possess oil and gas resources can increase trade barriers against the members of the regional content policy.

6.8.3 Damage of Infant Industries

A regional content policy protects the infant industries from the international market participants. But in a region where the development of countries varies, it is difficult to protect infant industries from regional powerhouses that will remove the infant firms from the value chain.

6.9 The East African Community

6.9.1 Introduction

Regional content policies are also productive development policies that increase economies of scale, bring about cooperation as well as increase the competitive advantage of the region. Regional content policies require a deeper level of integration in countries that adopt the policy for the oil and gas industry. Though the policy may be confused as a form of further economic regional integration, the policy does require facets similar to those of regional integration to succeed, such as geographic proximity, cultural, historical and ideological similarities, complementary economic linkages and a common language.[44] All these aspects are needed for the successful implementation of regional content policy.

The East African Community (EAC) is one of the first Regional Economic Arrangements in Sub-Saharan Africa. The EAC is made up of Kenya, South Sudan, Uganda, Tanzania, Rwanda and Burundi.[45] The

---

[43] South Sudan also has oil and gas resources but the political instability and security concerns make the country not be a viable member for the regional content policy.

[44] Kitson, M., Martin, R. and Tyler, P., 2004. Regional Competitiveness: An Elusive Yet Key Concept? *Regional Studies*, 38(9), pp. 991–999.

[45] Booth, D., Cammack, D., Kibua, T., Kweka, J. and Rudaheranwa, N., 2007. East African Integration: How Can It Contribute to East African Development? Overseas Development Institute Project Briefing.

EAC member states have several features in common such as, they all have a common history, language, culture and infrastructure.[46] These aspects have given East Africa an advantage to create a unique framework for regional cooperation and integration. The EAC the region has enjoyed increased trade, stretched its markets, attracted more FDI especially with the recent extractive discoveries, and increased bargaining power, strengthened security and conflict resolution in the region, as well as the free movement of people across the region.

The East African Community is the regional bloc that the book intends to attach the regional content policy. For the adequate application of the regional content policy, the history of the East African Community must be depicted in this book, to show whether the three East African countries selected namely Tanzania, Kenya and Uganda have the capability of adopting and implementing such a policy. The East African Community has ever collapsed due to various reasons that will be illustrated below. The case study intends to show the reasons that led to the collapse to ascertain that such an occurrence cannot happen again with the new re-invented East African Community. For local content policies to be effective and efficient in their application there some factors that must be present, these will be shown in the dominance of the new East African Community.

6.9.2 History of the East African Community

In East Africa, cooperation between Tanzania, Kenya and Uganda goes back to the pre-independence period. These countries having close geographic proximity and have traded across their borders for numerous years. They also shared a common colonial administration under the British rule.[47] In 1967 these three East African countries created the East African Community (EAC), which only lasted till 1977. After the collapse of the "first EAC", the East African countries negotiated and concluded the East African Community Mediation Agreement of

[46]Ibid.

[47]Kabananiye, A., 2011. Regional Integration and Sustainable Growth in Sub-Sahara Africa: A Case Study of the East Africa Community (Doctoral dissertation, Northumbria University).

1984.[48] Through this arrangement, the countries distributed among themselves the assets and liabilities of the obsolete Community, which marked the end of the economic bloc. The Mediation Agreement had left room for reconciliation and re-cooperation in the future.[49]

The re-cooperation Agreement in 1984 created room for the re-establishment of the East Africa Community in 1999.[50] The Treaty for the establishment of the new "East African Community" was signed by the Heads of State of the member states on November 30, 1999, in Arusha, Tanzania and came into force on 7th July 2000.[51] The Heads of State formally and officially launched the East African Community on 15th January 2001.[52] The main goal of the EAC as envisaged in its establishing Treaty is to broaden and deepen cooperation among member States in political, economic, social and cultural fields, research and technology, defence, security and legal and judicial affairs for their mutual benefits.[53] The vision adopted by the EAC is to have a prosperous, competitive, secure and politically united East Africa.

The mission, however, is to widen and deepen economic, political, social and cultural integration to improve the quality of life of the people of East Africa through increased competitiveness, value-added production, trade and investment. Emphasis is placed on accelerating development and building the capacity to compete effectively in the world economy. The first expansion of the Community was made in July 2007 when the countries of Rwanda and Burundi officially joined the

48 Ibid.

49 Hazlewood, A., 1979. The End of the East African Community: What are the Lessons for Regional Integration Schemes? *JCMS: Journal of Common Market Studies*, 18(1), pp. 40–58.

50 Mugomba, A.T., 1978. Regional Organisations and African Underdevelopment: The Collapse of the East African Community. *The Journal of Modern African Studies*, 16(2), pp. 261–272.

51 Hemingway, J., 2019. The East African Union: Regional integration in the East African community (Master's book).

52 Booth, D., Cammack, D., Kibua, T., Kweka, J. and Rudaheranwa, N., 2007. East African integration: How Can It Contribute to East African Development? Overseas Development Institute (ODI) Briefing.

53 Ibid.

Community.[54] Later in 2009 South Sudan also joined the EAC.[55] Therefore, the Community now has six member states, namely: Tanzania, Kenya, Uganda, Rwanda and Burundi and South Sudan.

6.9.3 Reasons for Collapse of the First East African Community

During the 1970s and 1980s, Africa was experiencing a nationalist wave. Many African countries wanted to gain independence, and this led to the ideologies that were different from those of their colonial administration.[56] Tanzania, under the leadership of Julius K. Nyerere, adopted a socialist ideology and played a pivotal role in fighting for independence in many African countries.[57] President Nyerere was among the "front-line" Presidents who led in providing diplomatic and military initiatives in Southern Africa.[58] Kenya was, at the same time, committed to being a capitalist state and gaining close allies in the Western world. Jomo Kenyatta the Kenyan President at the time fully embraced the support gained from the United States of America in an attempt to secure political, economic and diplomatic favour. Uganda was being ruled by Idi Amin, who was a notorious dictator at the time. The unpredictability of Amin led to so many ideological tumbles that Amin created more confusion within the region. The East African Community at the time could not survive this difference in ideology. Furthermore, Kenya and Tanzania

---

[54] https://www.voanews.com/a/a-13-2007-07-12-voa37/406029.html. High Expectations as Rwanda and Burundi Join East African Community.

[55] https://allafrica.com/view/group/main/main/id/00013706.html. South Sudan Ready to Join East African Community.

[56] Ravenhill, J., 1979. Regional Integration and Development in Africa: Lessons from the East African Community. *Journal of Commonwealth & Comparative Politics*, 17(3), pp. 227–246.

[57] Evans, M., 1986. The Front-line States South Africa and Southern African Security: Military Prospects and Perspectives.

[58] During the nationalistic wave of independence in Africa, the front-line states were states that were spearheading the independence movement as well as freedom from the colonial rulers. Such countries were Tanzania, Ghana, Zambia, Zimbabwe Egypt etc.

developed an estranged relationship. The region refused to recognise Amin's military regime in Uganda.[59]

Kenya being a capitalist state, did not want to carry the burden of the poorer countries like Tanzania that had been poor due to their socialist policies such as the Ujamaa policy. This made the East African countries gain different levels of economic development. Uganda and Tanzania resorted to depending on Kenya for manufactured goods because Kenya had been more industrious.[60] As a result, Kenya benefited the lion's share of the wide market. This situation left Uganda and Tanzania to think of establishing industries of their own. Because of this weakness, the community collapsed in 1977, apart from the different ideologies at the time in East Africa. Idi Amin overthrew Milton Obote's government in 1971.[61] President Obote was then given asylum in Tanzania by President Nyerere, who believed that Amin's regime was an illegal regime.[62] Amin's coup led to political instability in East Africa and a political crisis between Uganda and Tanzania. The conflict between Uganda and Tanzania led to the countries closing borders between themselves, suspension of internal flights by the East African Airways.[63] The animosity between Amin and Nyerere led to military engagements, and this complicated the operational functions of the East African Community. Furthermore, Amin declined to sign the EAC Appropriations Bill.[64] Tanzania reacted by denying Ugandans access to the EAC headquarters in Arusha, Tanzania. This led to the demise of the EAC in the region.

59 Ravenhill, F.J., 1974. Military Rule in Uganda: The Politics of Survival. *African Studies Review*, 17(1), pp. 229–226.

60 Mugomba, A.T., 1978. Regional Organisations and African Underdevelopment: The Collapse of the East African Community. *The Journal of Modern African Studies*, 16(2), pp. 261–272.

61 Ochen, E.A., 2017. The Military in Uganda: War, Peace and Support for Democratic Consolidation. In *African Peace Militaries* (pp. 54–78). Routledge.

62 Ibid.

63 Ambrose, K.W., 2010. Democracy and Development in Uganda. *OIDA International Journal of Sustainable Development*, 2(4), pp. 95–104.

64 Mugomba, A.T., 1978. Regional Organisations and African Underdevelopment: The Collapse of the East African Community. *The Journal of Modern African Studies*, 16(2), pp. 261–272.

6.9.4 Dominance of the New East Africa Community

East Africa' Infrastructure

It must first be established if the new EAC is susceptible to collapse like the EAC of the 1970s to adopt a regional content policy. The current EAC has been able to be a vehicle for economic and political cooperation in numerous aspects within the region. The East African Region has currently developed shared infrastructure, both soft and hard infrastructure. The Lamu Port and Lamu-Southern Sudan-Ethiopia Transport Corridor, a major infrastructural project within the East African region that is undertaken by members of the EAC.[65]

This project will involve road networks, oil pipelines, oil refinery, three airports, three resort cities (Lamu, Isiolo and Lake Turkana) and a standard gauge railway running through the capitals of Ethiopia and South Sudan.[66] Furthermore the East African Crude Oil Pipeline, an oil pipeline running 1443 km from Hoima, Uganda to Port Tanga in Tanzania with an investment of almost USD 3.5 Billion.[67] East African countries have reached an agreement to meet the demands of the power sector. This was done by establishing the Eastern Africa Power Pool (EAPP).[68] This is an organisation established in 2005 by countries within the region, namely Tanzania, Uganda, Libya, Democratic Republic of Congo (DRC), Egypt, Ethiopia, Kenya Rwanda and South Sudan. The EAPP was created with the sole mandate to facilitate access to the electric power supply to all these countries through regional power connections.[69] Such infrastructural projects within the East African region make it unlikely that the EAC can collapse like in the 1970s since the region now has more to lose and is deeper in integration.

---

[65] https://www.standardmedia.co.ke/business/article/2001313841/mega-projects-minting-millionaires-in-villages. Mega projects minting millionaires in villages.

[66] Ibid.

[67] https://www.theeastafrican.co.ke/business/Uganda-Tanzania-crude-pipeline-talks-continue/2560-4962886-33aijvz/index.html.Uganda-Tanzania Crude Pipeline Talks Continue.

[68] https://www.theeastafrican.co.ke/business/Phase-One-of-regional-power-pool-project-set-for-completion/2560-4699790-g706gj/index.html. Phase One of Regional Power Pool Project Set for Completion by Mid-2019.

[69] Ibid.

The Common Market

The common market of East Africa is the biggest achievement of the new EAC. The main objective of EAC's Common Market is to operate as a single market with a unified legal and fiscal framework, free movement of labour, capital, goods and services.[70] The EAC Common Market Protocol was established in July 2010.[71] The private sector has also supported the common market initiative by establishing the East African Commodities Exchange (EAX).[72] The EAX which is based in Kigali Rwanda was launched in Switzerland in January 2013. This platform links small scale farmers to sophisticated financing models while providing competitive prices for their agricultural produce through the regional commodities exchange. It is also notable that in 2013, the East African Securities Regulatory Authorities group, the umbrella body of regional capital markets regulators, launched a four-year strategic plan for the harmonisation of a capital markets regulatory and legal framework within the EAC. The ultimate objective is the complete integration of East Africa's capital markets. In addition, there is also a free movement of labour between Kenya and Rwanda. The two countries have removed work permit fees for all EAC citizens. The EAC has also harmonised immigration procedures at all border crossings.

Growth Prospects

East Africa has declared commercially viable discoveries of oil and gas in the region have raised the prospect of rapid economic growth. Nonetheless, even though the countries in the EAC are comparatively smaller economies to the West African economies. EAC integration has given

[70]Gastorn, K., and Masinde, W. 2017. The EAC Common Market. In E. Ugirashebuja, J. Ruhangisa, T. Ottervanger and A. Cuyvers (Eds.), *East African Community Law: Institutional, Substantive and Comparative EU Aspects* (pp. 285–292). Leiden; Boston: Brill. Retrieved from http://www.jstor.org/stable/10.1163/j.ctt1w76vj2.21.

[71]Basnett, Y., 2013. Labour Mobility in East Africa: An Analysis of the East African Community's Common Market and the Free Movement of Workers. *Development Policy Review*, 31(2), pp. 131–148.

[72]Ibid.

prospects for economic competitiveness that is yet to be achieved by its Southern and West African counterparts.[73] Due to the lucrative resources and the increased cross border investments and the integration of oil-rich countries as South- Sudan and Somalia that is considering joining the EAC. The region has the political will to sustain and develop further integration in the EAC. Substantial initiatives have been taken towards realising the EAC's objectives towards the adoption of an effective Customs Union and Monetary Union as well as establishing roots for the formation of a political federation.[74] With Kenya having discovered 700 million barrels of crude oil in Turkana, Uganda also discovered 6.5 billion barrels of crude oil in the Albertine region and Tanzania discovered 2.17 tcf of gas deposits in the Southern region of Mtwara and Lindi.[75] The EAC has been seen as a measure of cultivating policies for regional development within the region. The EAC cannot collapse due to the political will and the newly discovered extractive resources in this regard.

Political Stability

Unlike the 1970s, when Africa had developed a nationalistic wave towards fighting for independence. In today's post-independence era, the main enemy of East African countries is not a colonial power but poverty. East African countries have moved from developing political ideologies such as communism, revolutionary front-line countries, and socialism to economic development policies that will eradicate poverty and drive their economies into middle-income economies.[76] All East African countries are progressive democracies. The principles of good governance are yet to be fulfilled, but the political climate is

[73]Chingarande, A., Mzumara, M. and Karambakuwa, R., 2013. Comparative Advantage and Economic PERFORMANCE of East African Community (EAC) Member States. *Journal of Economics*, 4(1), pp. 39–46.

[74]Ibid.

[75]Anderson, D.M. and Browne, A.J., 2011. The Politics of Oil in Eastern Africa. *Journal of Eastern African Studies*, 5(2), pp. 369–410.

[76]https://www.brookings.edu/blog/future-development/2018/09/25/resource-rich-africa-can-boost-growth-to-reduce-poverty/. Resource-Rich Africa Can Boost Growth to Reduce Poverty.

stable compared to the 1970 era. The countries have moved from the euphoria of independence. Now believe in convergence and regionalism as the vehicle to sustainable economic development. East African countries have entered many regional agreements such as Southern African Development Corporation (SADC), the Common Market for Eastern and Southern Africa (COMESA), the Intergovernmental Authority on Development (IGAD), the East African Community (EAC) and lastly the African Continental Free Trade Area (Acfta).[77] Though overlapping memberships can be seen as a weakness, this shows that the East African countries have the political will of developing regional initiatives as vehicles of driving the economies towards economic development. Hence a regional content policy can be created in East Africa between Kenya, Uganda and Tanzania.

6.10 Reasons for Adopting a Regional Content Policy

East African countries have multiple policies, laws and regulations that are focused on increasing local capacity in East Africa. To enhance the quality of life and sustainably develop the economy. A regional content policy is being implemented as an umbrella policy that will harmonise all these regulatory requirements between the different regulatory bodies in the region. The African Economic Outlook states that "enhanced regional integration through the East African Community and the Common Market for Eastern and Southern Africa (COMESA) and potential exploitation of the oil and gas discoveries in Uganda, Kenya, Tanzania, have offered growth opportunities. Besides, urbanization and information and communications technology (ICT) development can support industrialization and structural transformation".[78] The implementation of a regional content policy can be drawn as a lesson from the

[77]Ntara, C.K., 2016. African Trading Blocs and Economic Growth: A Critical Review of the Literature. *International Journal of Developing and Emerging Economies*, 4(1), pp. 1–21.

[78]Outlook, A.E., 2018. African Economic Outlook 2018. Organization for Economic Cooperation and Development OECD, Development Centre, Paris, June.

Western African counterparts. Countries like Nigeria Ghana Cameroon and Angola have failed to change these lucrative resources into tangible economic outcomes.[79] Mainly due to countries acting alone in the regulation and integration of local firms thereof.

Though East Africa has experienced a bump in economic growth, this growth has been coming from non-resource-rich countries. The AfdB states that "growth was made possible by better macroeconomic management (not by the resource sectors), improved public sector investment, and increased household consumption thanks to the growing middle class".[80] A regional content policy will bring about structural transformation in the region. The regional policy must be supported by redistributive public policies tailored to the peculiar social-economic circumstances of the region. The regional content policy will diversify growth away from the traditional sectors in East Africa. The policy will create more jobs for the growing number of less-skilled and unemployed through the cross-country infrastructural projects, which are now limited to the domain of only the participating countries. A regional content policy invalidates the golden principle rule of local content policies in developing economies like Tanzania, Kenya and Uganda. The golden principle of local content entails that when the IOCs cannot acquire the goods and services within the boundaries of the local market, they can procure these goods and services from international markets through joint ventures between them and local firms.[81] However, the golden principle of local content gives room for the creation of joint ventures between local companies and foreign investors in the oil and gas industry.[82] It must be ascertained that in these resource-rich developing countries, there is no financial room or economic leverage to create joint ventures between the local firms and foreign investors. A regional content policy advocates first for regional joint ventures that will

[79] Brooks, S.M. and Kurtz, M.J., 2016. Oil and Democracy: Endogenous Natural Resources and the Political "Resource Curse". *International Organization*, 70(2), pp. 279–311.

[80] https://www.afdb.org/en/knowledge/macro-economics-policy-forecasting-and-research/.

[81] Ovadia, J.S., 2016. Local Content Policies and Petro-Development in Sub-Saharan Africa: A Comparative Analysis. *Resources Policy*, 49, pp. 20–30.

[82] Ablo, A.D., 2015. Local Content and Participation in Ghana's Oil and Gas Industry: Can Enterprise Development Make a Difference? *The Extractive Industries and Society*, 2(2), pp. 320–327.

be able to compete in the international market. East African countries have invested heavily in the development of human capacity, infrastructure, business, institutions, and technology and governance systems. Kenya, Uganda and Tanzania will use these as levers for transforming the economy through their local content regimes. Many of the roles in the oil and gas industry traditionally undertaken by foreigners have not been domesticated, so the local population have little opportunity to integrate the industry. This leaves the region perpetually dependent on foreigners and foreign expertise. The policies implemented to curb this dysfunctionality have no synergy in their application; this does not justify the investments made to integrate the local population in the oil and gas industry. Furthermore, a regional content policy can establish scalable formal enterprises, hence prioritising reforms that remove barriers to formal start-ups in East Africa.

6.11 Implementing a Regional Content Policy in East Africa

6.11.1 Approach for Adopting a Regional Content Policy

The regional content policy implements regional content requirements beyond national instruments. Thus the instruments to capture rents from the oil and gas industry such as fiscal and monetary measures are not applicable when adopting a regional content policy. Regional content concentrates on value addition that can be brought to national economies through national (using regional instruments) intervention in the oil and gas industry of a given region.

Traditionally local content requirements have been envisaged in different national policy instruments peculiar to the intentions of the policymakers of any given country.[83] These provisions can be found in oil

[83]Asiago, B.C., 2016. Fact or Fiction: Harmonising and Unifying Legal Principles of Local Content Requirements. *Journal of Energy & Natural Resources Law*, 34(3), pp. 337–360.

and gas agreements such as Production Sharing Agreements, or through legislative instruments such as regulations and decrees.[84]

But a regional content policy does not nullify the national content frameworks. In East Africa, the local content policies (national content) have been adopted throughout all sectors of the economy. The regional content policy will only regulate the oil and gas industry. Hence the requirements of the oil and gas local content provisions in the different national instruments will be found in the regional content policy, under the umbrella of the East African Community through the common market. A regional content policy must harmonise the interests of all stakeholders using an extensive stakeholder engagement in the East African region. The dialogue must ascertain the constraints, challenges and probable opportunities of all the stakeholders in East Africa. As well as create harmony and manage expectations of the different stakeholders in East Africa. Local content policies have always been considered unfavourable by International oil companies due to the divergent of interests between the foreign investors, governments and the general public.[85]

Furthermore, the regional content policy must have an analysis of the current local content legislation and policy in East Africa. This is because IOCs operate in an enclaved manner. Asiago states that "the majority of IOCs utilise established supply chain networks through their vertically integrated companies that give preference to global rather than local suppliers".[86] The UNCTAD report on global investments states that "as a cost-saving device IOCs frequently import labour, goods and services, which place local companies at a great disadvantage".[87] This has led resource-rich countries such as Tanzania, Kenya and Uganda

[84]Ibid.

[85]Ngoasong, M.Z., 2014. How International Oil and Gas Companies Respond to Local Content Policies in Petroleum-Producing Developing Countries: A Narrative Enquiry. *Energy Policy*, 73, pp. 471–479.

[86]Asiago, B.C., 2016. Fact or Fiction: Harmonising and Unifying Legal Principles of Local Content Requirements. *Journal of Energy & Natural Resources Law*, 34(3), pp. 337–360.

[87]UNCTAD, World Investment Report 2015: Reforming International Investment Governance. http://unctad.org/en/PublicationsLibrary/wir2015_overview_en.pdf. That is presuming that local companies have the competency to meet the requirements for the good or service in the first place.

adopt percentage-based requirements, legislative definitions on what local content should entail in a country and imposition contractual obligations to gain from there oil and gas resources. A regional content policy has to create a socio-economic balance and propel the potential of oil and gas industry into tangible sources of new employment opportunities and economic growth. The regional intervention implemented introduces a comprehensive but assertive regulatory framework. The regional content policy will consider the best international practise in implementing the regional content requirements in the East African region. It has already been established that countries such as Norway and Brazil already had suitable circumstances that could create the absorptive capacity in their economies for the large investment flows.[88]

Economies such as Norway and Brazil were technologically sophisticated with previous industries such as a thriving shipping industry that could be leveraged on in developing a skill for their oil and gas industry.[89] The advanced economies of Norway and Brazil had political stability, minimal corrupt tendencies in comparison to the East African countries.[90] Hence a regional content policy must not follow the national content approach that transplanted the local content approach from the design and requirements from these advanced economies into their poor weak economies. For example, In Kenya and Uganda, the local content policies advocate for local ownership of the oil and gas resources.[91] But for countries with a GDP less than $1000, such provisions are channels for corruption. A regional content policy's approach will not be based on ownership but rather on value addition, training provisions, provisions for R&D development and integration in the supply chains of the oil and gas industry. The underpinning implementation principle of a regional content policy are the three tenets of energy justice namely procedural justice, justice as recognition and distributional justice

[88] Rocha, F. and Ruiz, A.U., 2011. The Role of the Network Coordinator in the Attraction of Foreign Investment in R&D: The Case of the Brazilian Oil and Gas Industry. *Transnational Corporations*, 20(3), pp. 33–61.

[89] Heum, P., 2008. Local Content Development: Experiences from Oil and Gas Activities in Norway.

[90] Asiago, B.C., 2017. Norwegian Local Content Model a Viable Solution. *US-China L. Rev.*, 14, p. 471.

[91] Uganda and Kenya Local Content Regulations of 2018.

6.12 Regional Content Development in East Africa

6.12.1 Regional Content Development for Regional Suppliers

Regional supplier development can be enhanced by giving preference to the existing capabilities of the East African suppliers. Then enhancing these capabilities through regional competition and cooperation. This can be through one of the various ways such as availing access to finance, provide technology for SMEs and providing facilities for cluster formation etc.[92] All major programmes of work at all stages of the oil and gas life cycle will include regional capacity development and regional content strategies and plans

Procurement strategies to support regional content:

- Capacity development programmes in the East African region for the local employees and contractors in the region
- Support for regional training and research and development programmes.

The development of a regional content policy will facilitate the local firms and workforce in the region into becoming internationally competitive by implementing regional supplier development strategies such as[93]

- Unbundling of contracts to reflect the state and growth potential of the regional/ national market
- Market conditioning, including early notification of intended activities
- Preferential treatment of locals within the East African region

[92]Ablo, A.D., 2019. Actors, Networks and Assemblages: Local Content, Corruption and the Politics of SME's Participation in Ghana's Oil and Gas Industry. *International Development Planning Review*. January, pp. 1–22.

[93]Marcel, V., Tissot, R., Paul, A. and Omonbude, E., 2016. *A Local Content Decision Tree for Emerging Producers*. Chatham House, the Royal Institute of International Affairs: London, UK.

- Contract management with supplier performance improvement strategies
- Facilitating partnerships between private–public partnerships, creation of clusters, establishing R&D regional facilities, creating a regional hub for supplier development as well as establishing a regional supplier database so as to compete with the international suppliers
- Providing contract terms that facilitate access to competitive financing and opportunities to learn and grow, etc.
- Transparency of and accountability for decision-making in procurement.

Regional supply chain development will address regional development by investing in skills and training institutions, equipment and technology, infrastructure and facilities, facilitate capital aggregation.[94] Furthermore, a regional supply chain development will create linkages with other sectors and diversify the East African economy making the countries not to base their development plan on the oil and gas industry. Linkages will be enhanced along with the big four agenda that is sectors such as the manufacturing, construction, agri-business and tourism sectors.

The regional supplier strategies are in-line with the underpinning theory of energy justice. The unbundling of contracts creates more opportunities for local suppliers, hence serves justice as recognition. The market conditioning of having early notifications of intended activities serves the interests of procedural justice of having prior information. The transparency and accountability aspects of the strategy serve the distributional justice tenet of energy justice. The regional content policy can only be developed and succeed if the principles of energy justice have been attached to the policy framework.

[94] Ibid.

6.12.2 Regional Content Development for the Regional Workforce

A regional content framework provides provisions for the development of skills in three horizons. Firstly, enhance the existing capabilities through improving existing skills. A regional policy will have to adopt a risk management approach since East Africa is an immature market.[95] The first horizon will be policymakers will emphasise operational skills in the oil and gas sector. This will be done by designing training channels, building skills and enhancing managerial capabilities in the supporting industries.[96] The second horizon, a regional content framework, will assess the right level of intervention by building skills, determining the right regional (cross-country) partner. This will facilitate the transfer of knowledge and knowledge linkages all together. The third horizon is by creating regional centres of excellence by aligning government agenda and the regional agenda to enhance the regional capital market.

The Regional Content Policy and the African Continental Free Trade Area (AfCTA)

The AfCFTA is an ambitious agreement that brings together all the African countries to a single FTA. The main purpose of this comprehensive approach towards development is the creation of a common market for goods and services as well as allowing the free movement of persons between African nations.[97] AfCFTA emanates from the establishment of the Tripartite Free Trade Area (TFTA) that was concluded between

[95] Gelb, A., 2010. Economic Diversification in Resource Rich Countries. Centre for Global Development, pp. 1–23.

[96] http://www.ipieca.org/resources/good-practice/local-content-strategy-a-guidance-document-for-the-oil-and-gas-industry-1st-edition/. *Local Content Strategy: A Guidance Document for the Oil and Gas Industry* (1st edition).

[97] Parshotam, A., 2018. Can the African Continental Free Trade Area Offer a New Beginning for Trade in Africa?

COMESA, SADC and the EAC.[98] This Agreement places emphasis on developing regional industrial policies, bridging regional divisions as well as strengthening trade among different RECs in Africa. A regional content framework can leverage the features of the AfCFTA Agreement to implement provisions that enhance local participation in the regional oil and gas industry in East Africa.

The Regional Content Policy and the East Africa Treaty

The EAC Treaty creates a trade regime that facilitates cooperation and trade liberalisation as well as development. Due to the trade regime, the EAC created the Customs Union that provides for national treatment of goods and services from the member states. Thus a regional content policy will apply this provision to avoid any discrimination of foreign goods among the member states and the states adopting the regional content policy, namely Kenya, Uganda and Tanzania. Furthermore, the EAC Treaty establishes the Common Market in East Africa. This is known to be the biggest achievement of the regional economic bloc. The Common Market will facilitate the free movement of labour, goods, services, capital and the right of establishment. Hence the regional content policy can use the common market for transfer of goods and services within its framework and capacity.

Regional Content Policy and the Economic Community of West African States

West Africa has long suffered from the resource curse epidemic due to poor practises, corruption, civil and political instability and inadequate governance structures for natural resource management. At the same time, the ECOWAS region is blessed with natural resources among all Sub-Saharan African regions, both from non-renewable (oil; gas;

[98] Ibid.

uranium) and renewable (hydroelectric; solar and wind) sources.[99] With this in mind, ECOWAS built an ambitious regional energy policy, rooted in the 1975 ECOWAS founding treaty. The early goals were to foster collaboration and growth in all areas of economic activity, including natural resources; and to enhance the sub-regions collective resource autonomy.[100] In 1999, a regional agenda began to materialise slowly with the establishment in 2003 of the West African Power Pool (WAPP) and the ECOWAS Energy Protocol, which sought to foster long-term cooperation, increase complementarity and attract investment to facilitate regional energy trade in West Africa.[101]

The policy was further expanded with the creation of the West African Gas Pipeline (WAGP) in 2005 and the Renewable Energy and Energy Efficiency Centre (ECOWAS) in 2010.[102] Although substantial investments have been made through these regional initiatives to improve energy efficiency, almost all of them have faced major challenges in the form of repeated delays and/or increased transaction costs. Because the countries of West Africa lack unity and a shared interest in the governance of natural resources.[103] For example, in the energy sector; Nigeria, Ghana and to some extent Côte d'Ivoire are net exporters, largely through historical bilateral agreements, whereas most ECOWAS countries are largely dependent on electricity imports and fossil fuel.[104] A regional content policy would support regional infrastructure, which would also push production costs down and increase regional integration in the West African oil and gas industry.

[99] Fayomi, O. and Adeola, G.L., 2016. ECOWAS and Sub-Regional Integration in West Africa: An Appraisal. *Covenant University Journal of Politics and International Affairs*, 3(2).

[100] Ibid.

[101] Ouedraogo, N.S., 2013. Energy Consumption and Economic Growth: Evidence from the Economic Community of West African States (ECOWAS). *Energy Economics*, 36, pp. 637–647.

[102] Akinyemi, O., Efobi, U., Osabuohien, E. and Alege, P., 2019. Regional Integration and Energy Sustainability in Africa: Exploring the Challenges and Prospects for ECOWAS. *African Development Review*, 31(4), pp. 517–528.

[103] Ibid.

[104] Ouedraogo, N.S., 2013. Energy Consumption and Economic Growth: Evidence from the Economic Community of West African States (ECOWAS). *Energy Economics*, 36, pp. 637–647.

6.13 Summary

Regional integration has been an integral part of Sub-Saharan African policy and regulatory frameworks. Developing countries have pooled resources to meet several demands that can increase efficiency, industrialisation, as well as cutting down costs of production. The East African Community and the Economic Community of West African States, must collaborate to meet the sophisticated demands of the oil and gas industry. The cosmopolitan tenet of energy justice recognises people as world citizens. Giving regional governments the means to collaborate on certain regulatory aspects of the oil and gas industry. Individually, countries in Sub-Saharan Africa do not have the absorptive capacity to contain the huge investment inflow of FDI. Even if countries developed proper regulatory frameworks, their supplier base and local workforce are not readily available to supply the demands of the oil and gas industry at the right time, quantity, quality with timeous delivery. Countries like Norway and Brazil had a robust economy before the discovery of oil and gas resources. Industries such as shipping, fishing and manufacturing industries were well developed to provide linkages with the oil and gas sector. Unlike countries like Angola, Nigeria, Tanzania, Mozambique, Chad, Liberia, Ghana, Kenya, Uganda and Equatorial Guinea; whereby if they did not have the industrious capability then they had civil unrest leaving no room for development that could service the oil and gas industry. With the current wave of energy transition and the scare that Sub-Saharan Africa may remain with stranded assets. We must also understand that we might remain with "stranded-skills", it is important to collaborate and develop these skills faster before our natural resources lose value.

References

Abidin, M.Z., 2001. Competitive Industrialization with Natural Resource Abundance: Malaysia. In R.M Auty (Eds.), *Resource Abundance and Economic Development* (pp. 147–164). Oxford University Press.

Ablo, A.D., 2015. Local Content and Participation in Ghana's Oil and Gas Industry: Can Enterprise Development Make a Difference? *The Extractive Industries and Society*, 2(2), pp. 320–327.
Ablo, A.D., 2019. Actors, Networks and Assemblages: Local Content, Corruption and the Politics of SME's Participation in Ghana's Oil and Gas Industry. *International Development Planning Review*, January, pp. 1–22.
Akinyemi, O., Efobi, U., Osabuohien, E. and Alege, P., 2019. Regional Integration and Energy Sustainability in Africa: Exploring the Challenges and Prospects for ECOWAS. *African Development Review*, 31(4), pp. 517–528.
Ambrose, K.W., 2010. Democracy and Development in Uganda. *OIDA International Journal of Sustainable Development*, 2(4), pp. 95–104.
Anderson, D.M. and Browne, A.J., 2011. The Politics of Oil in Eastern Africa. *Journal of Eastern African Studies*, 5(2), pp. 369–410.
Asiago, B.C., 2016. Fact or Fiction: Harmonising and Unifying Legal Principles of Local Content Requirements. *Journal of Energy & Natural Resources Law*, 34(3), pp. 337–360.
Basnett, Y., 2013. Labour Mobility in East Africa: An Analysis of the East African Community's Common Market and the Free Movement of Workers. *Development Policy Review*, 31(2), pp. 131–148.
Booth, D., Cammack, D., Kibua, T., Kweka, J. and Rudaheranwa, N., 2007. East African Integration: How Can It Contribute to East African Development? Overseas Development Institute (ODI) Briefing.
Brooks, S.M. and Kurtz, M.J., 2016. Oil and Democracy: Endogenous Natural Resources and the Political "Resource Curse". *International Organization*, 70(2), pp. 279–311.
Chingarande, A., Mzumara, M. and Karambakuwa, R., 2013. Comparative Advantage and Economic Performance of East African Community (EAC) Member States. *Journal of Economics*, 4(1), pp. 39–46.
Emmanuel, R., 2016. Optimal Local Content Requirement Policies for Extractive Industries. *Resources Policy*, 50, pp. 244–252.
Evans, M., 1986. The Front-Line States South Africa and Southern African Security: Military Prospects and Perspectives.
Farooki, M. and Kaplinsky, R., 2014. Promoting Diversification in Resource-Rich Economies. *Mineral Economics*, 27(2–3), pp. 103–113.
Fayomi, O. and Adeola, G.L., 2016. ECOWAS and Sub-Regional Integration in West Africa: An Appraisal. *Covenant University Journal of Politics and International Affairs*, 3(2).
Gastorn, K. and Masinde, W., 2017. The EAC Common Market. In E. Ugirashebuja, J. Ruhangisa, T. Ottervanger and A. Cuyvers (Eds.), *East*

African Community Law: Institutional, Substantive and Comparative EU Aspects (pp. 285–292). LEIDEN; BOSTON: Brill. Retrieved from http://www.jstor.org/stable/10.1163/j.ctt1w76vj2.21.

Gelb, A., 2010. Economic Diversification in Resource-Rich Countries. Centre for Global Development (pp. 1–23). http://www.ipieca.org/resources/good-practice/local-content-strategy-a-guidance-document-for-the-oil-and-gas-industry-1st-edition/. *Local Content Strategy: A Guidance Document for the Oil and Gas Industry* (1st edition).

Hansen, M.W., 2014. From Enclave to Linkage Economies? A Review of the Literature on Linkages Between Extractive Multinational Corporations and Local Industry in Africa, DIIS Working Paper, no. 2, Copenhagen, Danish Institute for International Studies.

Hazlewood, A., 1979. The End of the East African Community: What Are the Lessons for Regional Integration Schemes? *JCMS: Journal of Common Market Studies*, 18(1), pp. 40–58.

Hemingway, J., 2019. *The East African Union: Regional Integration in the East African Community* (Master's book).

https://allafrica.com/view/group/main/main/id/00013706.html. South Sudan Ready to Join East African Community. https://www.afdb.org/en/knowledge/macro-economics-policy-forecasting-and-research/.

https://www.brookings.edu/blog/future-development/2018/09/25/resource-rich-africa-can-boost-growth-to-reduce-poverty/ Resource-rich Africa Can Boost Growth to Reduce Poverty.

https://www.standardmedia.co.ke/business/article/2001313841/mega-projects-minting-millionaires-in-villages. Megaprojects Minting Millionaires in Villages.

https://www.theeastafrican.co.ke/business/Phase-One-of-regional-power-pool-project-set-for-completion/2560-4699790-g706gj/index.html. Phase One of Regional Power Pool Project Set for Completion by Mid-2019.

https://www.theeastafrican.co.ke/business/Uganda-Tanzania-crude-pipeline-talks-continue/2560-4962886-33aijvz/index.html. Uganda-Tanzania Crude Pipeline Talks Continue.

https://www.voanews.com/a/a-13-2007-07-12-voa37/406029.html. High Expectations as Rwanda and Burundi Join East African Community.

Humphreys, M., Sachs, J.D. and Stiglitz, J.E. (Eds.), 2007. *Escaping the Resource Curse*. New York: Columbia University Press.

International Finance Corporation (IFC), 2015. The Art and Science of Benefit Sharing in the Natural Resource Sector, A discussion paper. International Finance Corporation, Washington, DC.

Jackson, J.H., 2000. The Perils of Globalization and the World Trading System. *Fordham Int'l LJ*, 24, p. 371.

Jacob, T. and Pedersen, R.H., 2018. New Resource Nationalism? Continuity and Change in Tanzania's Extractive Industries. *The Extractive Industries and Society*, 5(2), pp. 287–292.

Kabananiye, A., 2011. Regional Integration and Sustainable Growth in Sub-Sahara Africa: A Case Study of the East Africa Community (Doctoral dissertation, Northumbria University).

Lange, S. and Kinyondo, A., 2016. Resource Nationalism and Local Content in Tanzania: Experiences from Mining and Consequences for the Petroleum Sector. *The Extractive Industries and Society*, 3(4), pp. 1095–1104.

Marcel, V., Tissot, R., Paul, A. and Omonbude, E., 2016. *A Local Content Decision Tree for Emerging Producers*. Chatham House, the Royal Institute of International Affairs: London, UK.

Morrissey, O., 2012. FDI in Sub-Saharan Africa: Few Linkages, Fewer Spill-Overs. *European Journal of Development Research*, 24(1), pp. 26–31.

Mugomba, A.T., 1978. Regional Organisations and African Underdevelopment: The Collapse of the East African Community. *The Journal of Modern African Studies*, 16(2), pp. 261–272.

Ngoasong, M.Z., 2014. How International Oil and Gas Companies Respond to Local Content Policies in Petroleum-Producing Developing Countries: A Narrative Enquiry. *Energy Policy*, 73, pp. 471–479.

Ntara, C.K., 2016. African Trading Blocs and Economic Growth: A Critical Review of the Literature. *International Journal of Developing and Emerging Economies*, 4(1), pp. 1–21.

Ochen, E.A., 2017. The Military in Uganda: War, Peace and Support for Democratic Consolidation. In *African Peace Militaries* (pp. 54–78). Routledge.

Ouedraogo, N.S., 2013. Energy Consumption and Economic Growth: Evidence from the Economic Community of West African States (ECOWAS). *Energy Economics*, 36, pp. 637–647.

Ovadia, J.S., 2013. The Making of Oil-Backed Indigenous Capitalism in Nigeria. *New Political Economy*, 18(2), pp. 258–283.

Ovadia, J.S., 2016. Local Content Policies and Petro-Development in Sub-Saharan Africa: A Comparative Analysis. *Resources Policy*, 49, pp. 20–30.

Ovadia, J.S., 2017. Local Content in Tanzania's Gas and Minerals Sectors: Who Regulates? CMI Brief.

Parshotam, A., 2018. Can the African Continental Free Trade Area Offer a New Beginning for Trade in Africa?

Ravenhill, F.J., 1974. Military Rule in Uganda: The Politics of Survival. *African Studies Review*, 17(1), pp. 229–226.
Ravenhill, J., 1979. Regional Integration and Development in Africa: lessons from the East African Community. *Journal of Commonwealth & Comparative Politics*, 17(3), pp. 227–246.
Rocha, F. and Ruiz, A.U., 2011. The Role of the Network Coordinator in the Attraction of Foreign Investment in R&D: The Case of the Brazilian Oil and Gas Industry. *Transnational Corporations*, 20(3), pp. 33–61.
Uganda and Kenya Local Content Regulations of 2018.
UNCTAD, World Investment Report 2015: Reforming International Investment Governance. http://unctad.org/en/PublicationsLibrary/wir2015_overview_en.pdf. That is presuming that local companies have the competency to meet the requirements for the good or service in the first place.

7

Regional Content Policy Implications

7.1 Introduction

The oil and gas industry in developing countries and other developed countries has pushed for the adoption of local content policies. But the requirements of these policies have evolved from other production development policies, such as import substitution policies, indigenisation policies and nationalist resource policies, which are all means to achieve the objectives of local content policies. This chapter discusses the policy implications, harmonisation of national content requirements to create a regional content policy—the role of a regional content policy in creating regional linkages and regional value addition mechanisms. As well as the schematics of the regional content stakeholder engagement and lastly, the chapter discusses the potential international impact and issues in global value chains. This chapter examines the policy implications of regional content policies implementation in the East and/or the West African Community, it then explains the effects of harmonising different national content policies as well as the possibilities of creating a regional hub under the regional content policy. The chapter further explains the policy implications of regional content policy under international agreements such as the General Agreement on Tariffs and Trade. It

R. S. Muhongo, *Energy Justice*, Energy, Climate and the Environment,
https://doi.org/10.1007/978-3-030-61338-9_7

also explains the effects of a regional content policy in attaining sustainable development goals. Furthermore, the chapter provides an analysis of regional content metrics. Lastly, it explores the application of regional content policies in different spheres of local content governance, namely the macro level, meso-level and micro level of local content governance.

We also discuss the regional content literature within stakeholder relations from all spheres of governance within an oil and gas framework. The chapter brings about the implications of implementing regional content policies in any oil and gas industry. Through the case study methodology, the cross-country analysis of local content policies paved the way for the understanding of the regional content implications in a regions' oil and gas industry. The case study in this book navigates through different pivotal aspects for local content development. Norway and Brazil as the ideal local content countries offer lessons of transparency, good governance and the role of available infrastructure in nascent oil and gas industries. At the same time, mature Sub-Saharan oil and gas economies such as Nigeria and Angola show the role of politics in local content policies as well as the focus on local ownership than local skill and technological advancement. Furthermore, energy justice aligns with the recognition of different stakeholders at different levels of oil and gas governance. Energy justice provides a framework for regional content regimes playing a pivotal role in establishing a green economy since this assures a just transition of the skills and competence developed to crack through the enclave nature of the industry.

7.2 Regional Content Policy Implications

Regional content policies are yet to be implemented in any oil and gas region. The approach being adopted and implemented through the principles of regional integration must have some ramifications on a country to country basis and on the whole region. These implications range from national policies to international legal frameworks that affect the implementation of local content policies.

a. Policy Space to adopt a Regional Content Policy against the International Legal Framework:

In Sub-Saharan African countries have limitations in the types of developmental policies that they can implement. It has already been established that these policies foster linkages and other developmental schemes. However, policies such as those on local content have faced limitations due to international agreements. There is still space for the adoption of the regional content policy.[1] The WTO provides clear rules on what the regional content policy can implement and what the policy cannot implement. But the fundamental traits of regional content policies remain available for the policy to implement. Though the WTO rules do hinder the implementation of regional content policy, the following instruments are still permitted for the application of the requirements thereof.

- **Tariffs**: The use of tariffs is vital for the development of local capabilities as well as local participation through the development of local industries. The GATT does not prohibit the use of tariffs, but the legal instrument regulates the level of protection, by calling for binding requirements.[2] Countries like Kenya, Tanzania and Uganda have deficient levels of binding industrial tariffs, and when they do, the bound rates are higher than what is currently applied in practice. It should be noted that this is needed to maintain a certain degree of flexibility in case of any tariff increase if necessary to meet the objective of implementing the regional policy
- **Service Provisions:** GATTS regulate the implementation of service requirements in resource-rich countries. Unless these resource-rich

[1]The term 'policy space' appeared in literature in 2002 in UNCTAD documents. It was first used in official documents in the São Paulo Consensus of 2004. The Consensus defined the term policy space as 'the scope for domestic policies, especially in the areas of trade, investment and industrial development' which might be 'framed by international disciplines, commitments and global market considerations'.

[2]Ramdoo, I., 2016. Local Content, Trade and Investment: Is There Policy Space Left for Linkages Development in Resource Rich Countries? (No. 205). Discussion Paper.

countries are constrained through BITs and FTA, developing countries have sufficient policy space design to implement the following provisions in a regional content policy[3]:

I. Technology-related requirements, including technology transfer, financial contributions to support innovation and R&D;
II. Training requirements, including through specific spending requirements (such as a percentage of turnover or profits) or a number of people to be trained (e.g. number of scholarships to be financed on an annual basis);
III. Local employment requirements, including quotas and conditions for foreign labour; and,
IV. Ownership requirements, such as joint ventures, state participation or equity ratios and maximum foreign ownership.

- **Provisions for Subsidies:** Regional content policy will implement subsidies for the development of local companies and workforce. The ASCM prohibits the use of subsidies but offers some flexibilities for their implementation.[4] These are least-developed countries, countries with a GNP per capita of less than US$1000 per year which are listed in Annex VII to the ASCM[5]; and other developing countries. Other forms of subsidies permitted such as horizontal subsidies that apply to all economic sectors are exempt for the ASCM provided that these subsidies are not specific to a sector or an industry, i.e. if they are "neutral, economical in nature and horizontal in their application".[6] These include some of the following subsidies for infrastructure or corridor development, financial enticements, such as insurance or guarantees, credit finance, for example to SMEs, irrespective of the economic

[3]ASCM stipulates "objective criteria or conditions, as used herein, mean criteria or conditions which are neutral, which do not favour certain enterprises over others, and which are economic in nature and horizontal in application, such as the number of employees or the size of enterprise.

[4]GIZ, 2016. Space for Local Content Policies and Strategies. A Crucial Time to Revisit an Old Debate. Report prepared by the Columbia Centre for Sustainable Investment.

[5]Ibid.

[6]Ramdoo, I., 2016. Local Content, Trade and Investment: Is There Policy Space Left for Linkages Development in Resource Rich Countries? (No. 205). Discussion Paper.

sector, fiscal incentives and, not targeted to a definite industry. Subsidies targeting specific sectors are not prohibited. These provisions are not to be implemented if they may lead to adverse effects to another member of the WTO.[7] It should be noted that the country that claims to be adversely impacted bears the onus of proving so. In theory, states can provide sector-specific subsidies. Subsidies that are placed on services are not regulated under the GATS. Countries can provide financial support and other forms of incentives to support their local service suppliers. Regional policies supporting R&D and innovation are, however, considered as non-actionable.[8] A regional content policy can be designed to encourage local companies to innovate in new products and new production processes. The main implementation challenge for national content policies is not the policy space but the financial space for a single developing country to implement subsidies from their inception until their maturity. But with the adoption of regional content policy, the automatic policy implication is that policies will be implemented until they thrive due to the enormous financial background of the countries.

7.3 Harmonising National Content Requirements and Creating a Regional Hub

Resource-rich countries have implemented local content requirements through contractual instruments, judicial instruments and legislative instruments.[9] However, the implementation of these requirements has mostly been adopted within rigid regulatory frameworks in Sub-Saharan

[7]Article 5 of the ASCM states that "no Member shall cause, through the use of a subsidy, adverse effects to the Interests of other members, i.e. (a) injury to the domestic industry of another Member; (b) nullification or impairment of benefits accruing directly or indirectly to another Member under GAT 1994 in particular benefits accruing to concessions bound under Art. II of GATT 1994; and (c) serious prejudice to the interests of another member…".

[8]Article 8.2 of ASCM.

[9]Asiago, B.C., 2016. Fact or Fiction: Harmonising and Unifying Legal Principles of Local Content Requirements. *Journal of Energy & Natural Resources Law*, 34(3), pp. 337–360.

Africa. The rigidity of the local content requirements is ironic since the local populace lacks financial resources, technological awareness, skilled personnel, competitive goods as well as delivering services at the right price, quality and timeous delivery to suit the demands of the nascent oil and gas industry in the region. Furthermore, the oil and gas industry has a value chain characterised of diverse stages that represent separate subsectors and require different local integration mechanisms which make it incredibly unaffordable for non-competitive goods or services.[10] Arguably, at every stage of the value chain, the different countries adopting a regional content policy have set different priorities on how to integrate and participate in the oil and gas industry. Kenya, Angola and Nigeria have placed a focus on local ownership of the oil resources, while Uganda and Tanzania have emphasised local training of skills and furtherance of capabilities as well as local suppliers' development. Though the dependents of developing local content requirements are not only based on the design of the policy. The presence of the resources and the depletion policy adopted towards managing these resources are vital aspects when it comes to designing and developing a regional content policy. The policy implication is that the regional content policy will emphasise the development of skills, technological advancement and the implementation of the provisions for goods and services in East Africa. This is because if the regional content policy clings on very rigid requirements, it will hinder the growth of the oil and gas industry in East Africa by being unattractive to international oil companies.

East Africa is currently developing oil and gas industry mega-projects such as the East African Crude Oil Pipeline (EACOP) and the Lamu Port and Lamu-Southern-Ethiopia Transport Corridor (LAPSSET). These projects cross numerous borders. In West Africa, the ECOWAS proposed the development of a natural gas pipeline running throughout West Africa namely—Nigeria, Benin, Togo, Ghana and Cote d'Ivoire.[11] If

---

[10]Kazzazi, A. and Nourri, B., 2012. A Conceptual Model for Local Content Development in Petroleum Industry. *Management Science Letters*, 2, pp. 2165–2174.

[11]Mitchell, P., Santi, E. and Lichtenberg, Z., 2005, May. The Role of Socio-Political Risk Analysis in Developing Communications Strategies for Controversial Projects: The Case of the West Africa Gas Pipeline Project. In *CommNet Forum* (pp. 9–11). Washington: World Bank.

all countries adopt different levels of commitment of local participation as well as various mechanisms of supplier development within the region, this will lead to inefficiency and ineffective policies. Hence a regional policy harmonises the different requirements to enable a unified approach towards implementing and regulating regional content requirements.

7.3.1 Creating a Regional Hub

The international framework permits countries that have a GNP per capita less than $1000 to implement specific provisions for regional content policy.[12] Secondly, countries that have small proven reserves cannot implement adequate regional content requirements as the development of these policies may require time (that they don't have).[13] Kenya has a GNP per capita at $1, 991, Uganda's GNP per capita is at $600, and Tanzania's GNP per capita is at $900. According to these GNP margins, Kenya will not be permitted to implement certain requirements of the regional content policy. Secondly, Kenya's oil discoveries are 600 million barrels.[14] This makes Kenya a small oil-producing country, against its counterpart Uganda.

Moreover, due to the lower oil resources in Kenya, the government has abandoned the refinery construction project.[15] The viable investment of a refinery in a country is that the refinery must process at least 400,000 barrels of petroleum per day, Kenya can only plan to produce from the oilfields a maximum of 80,000 barrels per day.[16] It is therefore crucial for Kenya to lead in sharing oil facilities and capabilities with Uganda and Tanzania. Such cooperation would open

[12]Johnson, L. and Coleman, J., 2016. International Investment Law and the Extractive Industries Sector Briefing Note. Columbia Centre on Sustainable Investment.

[13]Nwapi, C., 2016. A Survey of the Literature on Local Content Policies in the Oil and Gas Industry in East Africa. SPP Research Paper (9/16).

[14]East Africa Economic Outlook, 2018. Macroeconomic Developments Manufacturing's Comparative Advantage and Competitiveness.

[15]Kenya says crude oil capacity insufficient for refinery. https://www.theeastafrican.co.ke/business/Kenya-says-crude-oil-capacity-insufficient-for-refinery/2560-4989226-q3uar1/index.html.

[16]Ibid.

opportunities for regional content development. Apart from the low reserves found in Kenya, the country also has a competitive economy than her counterparts—Kenya has more investors compared to Tanzania and Uganda.[17] The country has more talent with advanced technology compared to Tanzania and Uganda as well as a conducive business environment. The oil and gas industry involves global value chains. Kenya is involved in an international business model. Besides, regional headquarters and supply chain models that are motivated by regional integration are becoming a global trend,[18] The regional hub structure enables the IOCs to optimise resources, build capacity and foster quick technological advancements while maximising operational efficiencies.[19] Kenya functioning as a regional hub in the East African Community, will enable the other members adopting the policy to share resources and experiences.

7.4 Linkage and Regional Content Value Addition

Linkages in Sub-Saharan Africa can be enhanced by implementing a regional content policy. The development of linkages in a regional content policy is based on four aspects: the resources being extracted; specific country context; project life cycle and the ownership structure.[20] These are identified and briefly analysed below.

a. The Nature of the Resource being extracted:

[17]Taylor, S.D., 2018. Africa and Foreign Direct Investment. In *The Palgrave Handbook of African Politics, Governance and Development* (pp. 709–722). New York: Palgrave Macmillan.

[18]Malaysia on target to become regional oil and gas hub: MPRC https://www.nst.com.my/business/2017/09/278457/malaysia-target-become-regional-oil-and-gas-hub-mprc.

[19]Meet Malaysia: Investment Opportunities in Asia's Oil and Gas Hub Malaysia's Mature But Rapidly Growing Oil and Gas Sector Has Turned It into a Regional O&G Hub. http://www.mida.gov.my/home/42/pages/.

[20]Hansen, M.W., 2014. From Enclave to Linkage Economies? A Review of the Literature on Linkages between Extractive Multinational Corporations and Local Industry in Africa, DIIS Working Paper, 2014, No. 02. Copenhagen, Danish Institute for International Studies.

The nature of the natural resources determines what linkages regional content policy can foster. In East Africa, there is offshore oil and gas resources as well as onshore resources. These resources are found in Mtwara Southern gas-rich region of Tanzania, Turkana—the oil-rich region of Kenya, and the Albertine oil-rich region of Uganda. The offshore oil and gas resources require a high level of technical expertise. The development of linkages depends on the local know-how and skills base.[21] The regional content policy will increase the population of the neighbouring communities of the oil and gas resources with vast talent from all over East Africa. This develops linkages in the given communities due to the fiscal measures implemented, leading to fiscal linkages. Also, the money that will be spent in these areas will bring about consumption linkages while the presence of people will develop the infrastructure in these communities leading to spatial linkages. Furthermore, the construction of the EACOP and Lapsett project in Kenya, Uganda and Tanzania brings about an opportunity to enhance upstream linkages as an implication of the regional content policy.

b. The Country Context:

The regional policy development definitely depends on the characteristics of the countries that are adopting and implementing the policy. A regional content policy may bring about upstream, downstream, horizontal knowledge and technological linkages dependent on the nature and context of the countries.[22] Amendolagine states that "the lack of linkages observed in many African countries could, in part, be attributable to the recent nature of the extractive FDI-based development" [23] Sub-Saharan Africa has an immense knowledge and expertise gap with regards to oil and gas resource development since it is a new industry in these countries. A regional content enhances the chances of linkage creation by pooling talent and equally redistributing it among

[21] Ibid.

[22] Amendolaigne, V., Boly, A., Coniglio, N. D. and Prota, F., 2013. FDI and Local Linkages in Developing Countries: Evidence from Sub-Saharan Africa. *World Development*, 50, pp. 41–56.

[23] Ibid.

the given countries. This broadens the probability of regional linkage formation from the bottom- up.

c. The Life Cycle of the Project:

The development of linkages is minimal in the exploration phase of the industry since this phase requires more specialised skills. Yet, Sub-Saharan African countries individually do not possess these skills. The construction phase has high opportunities for linkages development.[24] Implementing a regional policy entails enhancing these linkages even further beyond the advantages to be gained in national content policy. Dangote's refinery in Nigeria, employs over 40,000 people and upon completion, the refinery will employ over 100,000 local Nigerians. The refinery is set to 650,000 barrels per day (BPD) integrated refinery and petrochemical[25]—another large project, the Chad-–Cameroon pipeline and the West African Gas Pipeline. In East Africa, the construction of the EACOP is creating over 15,000 jobs between Uganda and Tanzania.[26] The regional content policy advocates for more linkage development. Regional content can maximise consumption linkages by moving the large numbers of personnel throughout Sub-Saharan Africa enabled by the Free Movement of People Protocols. The policy also creates linkages in the production phase of the industry. By creating clusters and pooling capital from the Sub-Saharan African suppliers, the local business can integrate the enclaved oil and gas industry.

d. The Ownership Structure of the Oil and Gas Companies:

Morris et al. (2012) state that "foreign ownership might be less conducive to linkage creation than domestic ownership for cultural and

[24] Kaplinsky, R., 2011. Commodities for Industrial Development: Making Linkages Work. Working Paper 01/2011, United Nations Industrial Development Organization.

[25] Adetayo, S., 2020. The Ethics of State Capture: Dangote and the Nigerian State. In *The Palgrave Handbook of African Social Ethics* (pp. 371–388). Cham: Palgrave Macmillan.

[26] Graham, E. and Ovadia, J.S., 2019. Oil Exploration and Production in Sub-Saharan Africa, 1990–Present: Trends and Developments. The Extractive Industries and Society.

nationalistic reasons".[27] A regional content policy creates a competitive and comparative advantage among the states that adopt the policy. This brings about a mandate to spur domestic economic activities through upstream, downstream, spatial and knowledge linkages. A regional content policy integrates local oil and gas companies into the oil and gas value chain. The policy will lead the enclaved oil and gas companies to move towards greater outsourcing of their non-core competencies, integrate community-oriented and corporate social responsibility will be adopted to minimise risk.[28] Through the competitive nature of the policy, local companies will become extractive champions, enabling Sub-Saharan African governments to strengthen their bargaining power for the development of the sector.

7.5 Regional Content Value Addition

In resource-rich countries, the emphasis is being placed on value addition to create additional employment opportunities and the need to diversify the economy. Value addition is a crucial feature in the objectives and principles behind adopting a regional content policy. Ramdoo (2015) states that "while some resource-rich countries have provided incentives for and co-financed the construction of refineries (for example Angola and Uganda), others have used taxes and restrictions on exports to encourage downstream processing – (Tanzania)".[29] It has not been possible for national content interventions to draw advantages from refining activities in the oil and gas sector. Value addition is capital intensive. As such, it calls for adequate infrastructure and faces highly

[27] Morris, M., Kaplinsky, R. and Kaplan, D., 2012. One Thing Leads to Another—Commodities, Linkages and Industrial Development—A Conceptual Overview, MMCP Discussion Paper, Milton Keynes and Cape Town, The Open University and University of Cape Town.

[28] Morris, M., Kaplinsky, R. and Kaplan, D., 2012. One Thing Leads to Another—Commodities, Linkages and Industrial Development—A Conceptual Overview, MMCP Discussion Paper, Milton Keynes and Cape Town, The Open University and University of Cape Town.

[29] Ramdoo, I., 2015. Resource-Based Industrialisation in Africa Optimising Linkages and Value Chains in the Extractive Sector, Discussion Paper, No. 179, European Centre for Development Policy Management.

competitive companies from the global market.[30] Value addition has been challenging to set up in Sub-Saharan Africa due to the countries not having adequate infrastructure and the capital to move further downstream in this sector. Regional content can facilitate value addition by making the proximity to the oil and gas resources a comparative advantage. Hausmann states that "raw commodities play a small role in determining comparative advantages, even for high volume products that are associated with higher transport costs".[31] Since raw materials cannot create the comparative advantage thereof, the governments in East Africa can turn to policy instruments to enable the competitive advantage of these resources. That brings about strong positions to promote value addition initiatives. The policy justifies this by creating a competitive advantage in the long run and protecting the industry in the short run, which can be traced to the infant industry argument.

Value addition is also enabled by the creation of lateral linkages in the oil and gas sector. This is by deepening integration while increasing collaboration through the regional development policy. The degrees of integration must be tactical, operational and strategic among the regional content policy adopting countries. The collaboration among the different companies, state agencies and other actors brings about the adaptability of different abilities in the nascent industry. As Kaplinsky states that "the probability that a country will develop the capability to be good at producing one good is related to its installed capacity in the production of other similar, or nearby goods for which the currently existing productive capabilities can be easily adapted".[32]

---

[30] Ibid.

[31] Hausmann, R., Klinger, B. and Lawrence, R., 2008. *Examining Beneficiation.* Cambridge: Harvard University.

[32] Kaplinsky, R., 2011. Commodities for Industrial Development: Making Linkages Work, Working Paper 01/2011, UNIDO.

7.6 Regional Content and Stakeholder Engagement

The regional content policy has numerous stakeholders, but the stakeholders are divided into three spheres of the regional economic structure. These are the macroeconomic, the meso-economic and the microeconomic spheres which are detailed in turn below. These are important because they reveal to what extent regional content policy can be implementable and how many stakeholders are affected by the policy

a. The Macroeconomic stakeholders and conditions

The regional policy framework at the macro scale will involve regional and state actors, ministries, and national oil companies' etc. These stakeholders have the following mandate in adopting the regional content policy: They must have specified systems for monitoring company obligations; The government and regional stakeholders need to have clearly defined legal obligations for the companies; create uniform frameworks for all oil and gas companies in the region; establish, oversight institutions with relevant capacities; and monitor and enforce compliance to the regional policy commitments.[33] The macroeconomic stakeholders must utilise both the regional policies and the national policies to develop the regional content policy and enhance their competitive and comparative advantage in the oil and gas industry. These include the regional instruments such as the Common Market Protocol found in East African Treaty, the Southern African Development Corporation and the African Continental Free Trade Area,[34] the cooperation in investment and industrial development strategy.[35] Regional trade regimes, cooperation, trade liberalisation, development policy, cooperation in

[33]Oloruntoba, S.O., 2019. Innovation and Regional Integration in Africa: Exploring Theory and Praxis for Socio Economic Development. In *Innovation, Regional Integration, and Development in Africa* (pp. 29–41). Cham: Springer.

[34]African Union, 2019. Protocol on Relations Between the African Union (AU) and the Regional Economic Communities (RECs). https://archives.au.int/bitstream/handle/123456789/1621/Protocol_Relations_AU_RECs_E.pdf?sequence=1&isAllowed=y.

[35]Ibid.

infrastructure and services, cooperation in the development of human resources, science and technology, free movement of persons, labour, services, right of establishment and residence, enhancing the role of women in socio-economic development, cooperation in political matters, private sector and civil society and the policy on relations with other regional and international organisations and development partners.[36] The national polices, and legal framework must intertwine with these regional policies under regional economic blocs as an implication of developing the regional content policy. The policy will trigger economic growth and enhance regional competitive and comparative advantage through supporting entrepreneurship and resource development among the Sub-Saharan African countries.

b. The Meso-Economic Stakeholders:

The second sphere of stakeholders lays in the meso-economic sphere. This sphere of regional content development is linked to the specified location of the oil and gas resources in Sub-Saharan Africa concerning natural resources, technical, economic and social infrastructural aspects and policy adopted by the local authorities of the communities next to the oil and gas operations. The oil and gas resources determine the effectiveness of the regional content policy in the region. The adjacent communities of the oil and gas operations possess the raw materials and the resources before processing and adding value to these resources. Remote regions like Mtwara in the gas-rich region of Tanzania, the oil-rich Albertine region of and the refinery being constructed in Hoima and the oil-rich region in Turkana, Kenya, the Rovuma–Basin in Mozambique, the Niger Delta in Nigeria. These remote areas must not only offer public goods to the oil and gas industry but also gain from their resources. The policy should leverage the existing transporting networks, telecommunication networks as well as energy infrastructure for the inclusion of these communities into the oil and gas value chain.[37] The

[36] African Union, 2019. Protocol on Relations Between the African Union (AU) and the Regional Economic Communities (RECs).

[37] Aiginger, K. and Firgo, M., 2015. Regional Competitiveness Under New Perspectives.

policy must advocate for the creation of alternative sources of energy[38] since oil and gas resources are found in remote areas, with no electricity, water and other basic needs. The policy must enhance the use of alternative sources of energy as a temporary measure to facilitate the creation of linkages and as a quick measure to creating the local absorptive capacity.

The economic infrastructure refers to mainly the proximity of the oil and gas resources to the production components, the different suppliers and financial institutions. The policy aligns the surrounding communities to the regional bank networks, venture capital institutions, SMEs financial incentives. All these aspects are part of the policy implications of adopting a regional content policy. Furthermore, the policy will link the social infrastructure to the oil and gas resources. The policy determines the housing resources, health, and educational infrastructure. The marginalised communities of Sub-Saharan Africa have a direct impact of the oil and gas industry operations. This is why these communities must also have tangible development by creating good access to housing facilities that will result in developing regional competitiveness through increasing labour mobility in the region. Furthermore, hospitals, clinic and pharmacies all are vital determinants of increasing regional competitiveness as well as increasing the comparative advantage through the regional content policy. The policy also aligns all these factors with the commitments of the local authorities where these oil and gas operations take place. In Kenya, they have a federal system of governance—Kenya enacted the County Acts, which regulate the governance structure and function of the County governments.[39] In Uganda and Tanzania, the local governments still answer to the central government. But both the County Act and Local Government Act of these countries advocates for the proper governance, local inclusion and the sustainable development of their resources. The regional content policy facilitates local authorities to establish small business supporting mechanisms. Support CSOs and other skills and local supplier development providers, simplify business

[38]Huenteler, J., Niebuhr, C. and Schmidt, T.S., 2016. The Effect of Local and Global Learning on the Cost of Renewable Energy in Developing Countries. *Journal of Cleaner Production*, 128, pp. 6–21.

[39]Kramon, E. and Posner, D.N., 2011. Kenya's New Constitution. *Journal of Democracy*, 22(2), pp. 89–103.

processes and avail the social license to operate for both IOCs and other market participants coming from the region, develop subnational plans, align local content and subnational development plans, and also facilitate funding mechanisms for marginalised groups within these communities such as women, elderly and the disadvantaged.

c. The Micro-Economic Stakeholder

The micro-economic sphere of the policy implication of regional content in East Africa will entail developments for both enterprises and the local inhabitants. The stakeholders in the microeconomic sphere directly link with both the local entrepreneurs and the IOCs in the Sub-Saharan Africa region. The relationship between the regional policy and these market participants is determined by the following factors: The size of the entrepreneurs; the scope of the business activity in different localities; the economic situation and the pace of exploiting the resources. The IOCs in this sphere of regional content policy are expected to communicate opportunities, develop workforce and supplier development programmes, provide access to finance (cash, credit and equity), cooperate within the industry, support CSOs, produce regular, transparent, auditable reports, diagnose demand and supply capacities, develop and report on regional content strategy and invest in capability development. The local entrepreneurs do not have the financial and technological backing to develop to integrate the oil and gas value chain efficiently. The firms in these regions are characterised with low business activity accompanied by minimal strategies to change the situation thereof.[40] Fesshaie states that, "low public investment in technological capacities leads African suppliers into a vicious circle of low access to investment capital, low technological capabilities to start with, low incentives to adopt new technologies, and high risk that the market will not reward such investment".[41] The policy attempts to link the enterprise factors with the local inhabitants of the East African region. This is

[40]Anderson, D.M. and Browne, A.J., 2011. The Politics of Oil in Eastern Africa. *Journal of Eastern African Studies*, 5(2), pp. 369–410.

[41]Morris, M. and Fessehaie, J., 2014. The Industrialisation Challenge for Africa: Towards a Commodities Based Industrialisation Path. *Journal of African Trade*, 1(1), pp. 25–36.

through the common market protocol in East Africa that will facilitate the migration of capital. The policy will also improve cooperation between different stakeholders in the region. The regional content policy also addresses the mobility of qualified personnel from one region to the other. Making sure that there is no rent-seeking since there is a "bigger-pie" to share among these communities.

7.7 Regional Content and Potential International Impact and Issues

Regional content policies have an impact on different levels of energy law, namely the international level, national level and the community level. In the regional economic bloc sphere of which this would be the international level of energy law in this regard. The policy has an impact on increased transparency and accountability, this enhances the efficient utilisation of oil and gas resources in the region as well as increase the adherence of energy justice in the oil and gas governance framework.

a. Regional Content and EITI Initiatives

The Extractive Industries Transparency Initiative (EITI) is an international, multi-stakeholder initiative that enhances transparency and accountability in the oil and gas industry.[42] Through the disclosure of information by both the companies and the governments thereof. Only Tanzania in East Africa has an Act regulating EITI in the country.[43] Kenya and Uganda must first become candidate countries and then compliant countries.[44] The process of adopting EITI is as follows, and the states must issue an explicit public statement of their intentions to

[42]EITI, The EITI Standard, 2016, available at: https://eiti.org/document/standard.

[43]Sovacool, B.K., Walter, G., Van de Graaf, T. and Andrews, N., 2016. Energy Governance, Transnational Rules, and the Resource Curse: Exploring the Effectiveness of the Extractive Industries Transparency Initiative (EITI). *World Development*, 83, pp.179–192.

[44]Sturesson, A. and Zobel, T., 2015. The Extractive Industries Transparency Initiative (EITI) in Uganda: Who Will Take the Lead When the Government Falters? *The Extractive Industries and Society*, 2(1), pp. 33–45.

implement the EITI.[45] Then they must each appoint a senior individual to lead the implementation of the EITI in their respective countries. Uganda and Kenya have both to commit to working with civil society and companies to establish a multi-stakeholder group (MSG) to administer the implementation of the EITI. After all the countries adopting the regional content policy have implemented the EITI, this will benefit the region in numerous ways such as:

I. **Inform oversight**. The information availed in the EITI reports can help in the identification of governance weaknesses. This will assist the regional content policy stakeholders in monitoring the management of the oil and gas industry as well as develop recommendations for policy reform.[46]
II. **Inform sector law-making.** EITI reports create transparency in the allocation of oil and gas rights, the functioning of national oil companies, the efficiency of revenue collection, and shifts in production.[47]
III. **Ensure the government is working in the interests of the public.** The EITI initiatives can help the regional content policy ensure that the government is concluding a favourable Agreement for the country.[48]

7.8 Regional Content Metrics

Performance metrics are a central part of the oil and gas industry. Besides, for regional content to play an integral role in oil and gas regulation in Sub-Saharan Africa, the choice of metric measures for regional content

[45]Guidance to implementing the EITI Standard can be found on the EITI website at: https://eiti.org/guide.
[46]NRGI, From Reporting to Reform: Eleven Opportunities for Increasing EITI Impacts, 2015, available at: https://resourcegovernance.org/analysis-tools/publications/reporting-reform-eleven-opportunities-increasingeiti-impacts.
[47]Ibid.
[48]NRGI, From Reporting to Reform: Eleven Opportunities for Increasing EITI Impacts, 2015, available at: https://resourcegovernance.org/analysis-tools/publications/reporting-reform-eleven-opportunities-increasingeiti-impacts.

must be emphasised. Metrics adopted in the oil and gas sector come in three categories, and these are[49]:

a. The extent to which national citizens or local suppliers are capable of capturing expenditure by oil and gas companies on labour and goods and services;
b. The ability of a company, government, or supplier to grow its share of local content. This can be through provisions of building skills in the national workforce through training and education or investing in developing the capabilities and competitiveness of domestic supply chains; and
c. The metric for measuring the level of regional content development within a given region.

The regional content policy will concentrate on the supplier chain management as that is where most linkages can be attained in the oil and gas industry. With regard to the regional content in the workforce, the stipulations of the Production Sharing Agreements will be considered. The metric for measuring regional suppliers measures the expenditure that is spent on the regional suppliers by the IOCs and other market participants. The definitions used for the classification of what is spent on the local suppliers fundamentally includes one or a combination of the following criteria[50]:

a. The address in the vendor registration information
b. The address on a purchase order or invoice
c. The geographic location of the service being provided or the production of the good being supplied
d. The share of equity owned by national citizens (for example, greater than 25%, 50%, or 100%)
e. Whether the supplier is locally incorporated

---

[49]Warner, M. 2010. "Unpacking Local Content Metrics and Measurement: Local Content Solutions." Briefing No. 5, Local Content Solutions. http://www.localcontentsolutions.com/pdf/SolutionsSeries5.pdf.

[50]Tordo, S., Warner, M., Manzano, O.F. and Anouti, Y. 2013. *Local Content Policies in the Oil and Gas Sector*. Washington, DC: World Bank Study, World Bank.

f. Whether the supplier is tax registered in the country of operation, including for withholding tax purposes
g. Whether the supplier employs more than a specified percentage of nationals.

According to the World Bank, most of the expenditure on goods and services is done in companies that are owned by foreigners.[51] This creates minimal linkages, minimal local supplier development in the country and minimal workforce development in this regard. A regional content will act as a buffer towards this anomaly that causes market failure and information asymmetry in the region. The regional content policy will measure the expenditure of regional goods and services according to whether the services or goods purchased in the region are deemed to be in some way to be local. The reference towards classifying the expenditure of goods and services through a local supplier, will not serve the function of the policy for this nascent industry in the region, because the local population does not have the absorptive capacity to meet the demands of the oil and gas industry. The classification of local goods and services can be regarded as some form of rules of origin. Tordo states that "rules of origin were originally applied to determine whether a good or product was eligible for preferential treatment in the course of trade between nations or regional trading communities. These rules are now being applied by some regulators to formulate local content targets and reporting requirements".[52] This means that an IOC or any other oil and gas market participant will be able to measure regional content development only if there is value added in the region to provide a service or produce a product. The mere registration or the fact that the product may be owned by a local person or company within the region will not constitute as regional content development. This is because most local

[51] Ibid.
[52] Tordo, S., Warner, M., Manzano, O.F. and Anouti, Y., 2013.*Local Content Policies in the Oil and Gas Sector*. Washington, DC: World Bank Study, World Bank.

entrepreneurs are mere local importers; thus, no value addition is done within the region.[53]

The regional content policy emphasises the importance of value addition. This necessitates the need to measure value addition. The build-up method of calculating value addition will be applicable in this regard. This is because of the following main reasons: The regional content policy is applicable to one or more different countries; the policy affects more than one industry; furthermore, the policy is aligned with the East African industrialisation policy.[54] These factors are key in adopting the build-up method, as Tordo states "build-up methods aggregate piecemeal - the value added to the domestic economy from in-country industrial activity, that is, the value of the locally originating raw materials and components, the direct labour costs, direct overhead costs, internal transportation, and profit or mark-up. The build-up methods are thus generally more accurate but also more complex to piece together, requiring data drawn from a range of company functions: procurement, finance, human resources, sales, and so on".[55] The fact that a regional policy encompasses numerous stakeholders with different expectations, the accuracy of the build-up method makes it a credible calculating method for value addition in the region. The build-up method concentrates on in-country industrial activity, and the regional content policy emphasises on industrial policy to facilitate direct domestic investment making the build-up method more suitable mechanism for the policy. Metric for regional content development; would implement output metrics in reporting regional content development. Policymakers would need to measure the sustainability and competitiveness gained over time through assessing the supplier costs, delivery times, labour productivity,

[53]This affects the depth and breadth of linkages. Since most entrepreneurs in resource rich countries are indeed importer and re-sellers. This shows that there is no depth of linkages, because the manufacturers are not involved in the process.

[54]The East African Community Industrialisation Policy in Brief East African Community Opening new opportunities for growth and expansion of cross border manufacturing and upgrading of small and medium enterprises (SME). https://eacgermany.org/wp-content/uploads/2014/10/EAC-Industrialization-Policy-Brief.pdf.

[55]Tordo, S., Warner, M., Manzano, O. F. and Anouti, Y., 2013. *Local Content Policies in the Oil and Gas Sector*. Washington, DC: World Bank Study, World Bank.

quality of service or product, volume capacity, and HSE performance,[56] to evaluate the development of local suppliers.

7.9 Summary

Sub-Saharan African countries have pooled resources for several years. The aggregate effort in meeting specific challenges in Sub-Saharan Africa is fundamentally driven by politics and not industry. Cracking through the oil and gas enclave and create linkages to other sectors of the economy has been challenging. Because developing countries do not possess the financial or technological capability and expertise to transform their resources into vehicle of economic development. The regional content policy offers a unique possibility for natural resource management and governance. Creation of transporting corridors, creating regional hubs, creating regional technical institutions, increased accountability, increased transparency and increased stakeholder engagement are all possible with a regional content policy. The harmonisation of different policies attracts more investments into the region. Fuelling development and social welfare, facilitating to reach the Sustainable Development Goals (SDGs) that are important for any country's development.

References

Adetayo, S., 2020. The Ethics of State Capture: Dangote and the Nigerian State. In *The Palgrave Handbook of African Social Ethics* (pp. 371–388). Cham: Palgrave Macmillan.

Amendolaigne, V., Boly, A., Coniglio, N.D. and Prota, F., 2013. FDI and Local Linkages in Developing Countries: Evidence from Sub-Saharan Africa. *World Development*, 50, pp. 41–56.

[56] WBCSD (World Business Council for Sustainable Development), 2012. A Framework for Dialogue on National Market Participation and Competitiveness. Geneva: WBCSD. http://www.wbcsd.org/Pages/EDocument/EDocumentDetails.aspx?ID=14421&NoSearchContextKey=true

Asiago, B.C., 2016. Fact or Fiction: Harmonising and Unifying Legal Principles of Local Content Requirements. *Journal of Energy & Natural Resources Law*, 34(3), pp. 337–360.

Dialogue on National Market Participation and Competitiveness. Geneva: WBCSD.

East Africa Economic Outlook, 2018. Macroeconomic Developments Manufacturing's Comparative Advantage and Competitiveness.

GIZ, 2016. Space for Local Content Policies and Strategies. A Crucial Time to Revisit an Old Debate. Report prepared by the Columbia Centre for Sustainable Investment.

Graham, E. and Ovadia, J.S., 2019. Oil Exploration and Production in Sub-Saharan Africa, 1990–Present: Trends and Developments. The Extractive Industries and Society.

Guidance to implementing the EITI Standard can be found on the EITI website at https://eiti.org/guide.

Hansen, M.W., 2014. From Enclave to Linkage Economies? A Review of the Literature on Linkages Between Extractive Multinational Corporations and Local Industry in Africa, DIIS Working Paper, 2014, No. 02. Copenhagen, Danish Institute for International Studies.

Hausmann, R., Klinger, B. and Lawrence, R., 2008. *Examining Beneficiation.* Cambridge: Harvard University. http://www.wbcsd.org/Pages/EDocument/EDocumentDetails.aspx?ID=14421&NoSearchContextKey=truc.

Huenteler, J., Niebuhr, C. and Schmidt, T.S., 2016. The Effect of Local and Global Learning on the Cost of Renewable Energy in Developing Countries. *Journal of Cleaner Production*, 128, pp. 6–21.

Johnson, L. and Coleman, J., 2016. International Investment Law and the Extractive Industries Sector. Briefing Note. Columbia Centre on Sustainable Investment.

Kaplinsky, R., 2011. Commodities for Industrial Development: Making Linkages Work, Working Paper 01/2011, UNIDO.

Kazzazi, A. and Nourri, B., 2012. A Conceptual Model for Local Content Development in Petroleum Industry. *Management Science Letters*, 2165–2174.

Kenya Says Crude Oil Capacity Insufficient for Refinery. https://www.theeastafrican.co.ke/business/Kenya-says-crude-oil-capacity-insufficient-for-refinery/2560–4989226-q3uar1/index.html.

Kramon, E. and Posner, D.N., 2011. Kenya's New Constitution. *Journal of Democracy*, 22(2), pp. 89–103.

Malaysia on Target to Become a Regional Oil and Gas Hub: MPRC. https://www.nst.com.my/business/2017/09/278457/malaysia-target-become-regional-oil-and-gas-hub-mprc.

Meet Malaysia: Investment Opportunities in Asia's Oil and Gas Hub Malaysia's Mature But Rapidly Growing Oil and Gas Sector Has Turned It into a Regional O&G Hub. http://www.mida.gov.my/home/42/pages/.

Mitchell, P., Santi, E. and Lichtenberg, Z., 2005, May. The Role of Socio-Political Risk Analysis in Developing Communications Strategies for Controversial Projects: The Case of the West Africa Gas Pipeline Project. In *CommNet Forum* (pp. 9–11). Washington: World Bank.

Morris, M., Kaplinsky, R. and Kaplan, D., 2012. One Thing Leads to Another: Commodities, Linkages and Industrial Development—A Conceptual Overview, MMCP Discussion Paper, Milton Keynes and Cape Town, The Open University and University of Cape Town.

NRGI, From Reporting to Reform: Eleven Opportunities for Increasing EITI Impacts, 2015. Available at: https://resourcegovernance.org/analysis-tools/publications/reporting-reform-eleven-opportunities-increasingeiti-impacts.

Nwapi, C., 2016. A Survey of the Literature on Local Content Policies in the Oil and Gas Industry in East Africa. SPP Research Paper (9/16).

Oloruntoba, S.O., 2019. Innovation and Regional Integration in Africa: Exploring Theory and Praxis for Socio Economic Development. In *Innovation, Regional Integration, and Development in Africa* (pp. 29–41). Cham: Springer.

Ramdoo, I., 2015. Resource-Based Industrialisation in Africa Optimising Linkages and Value Chains in the Extractive Sector, Discussion Paper, No. 179, European Centre for Development Policy Management.

Ramdoo, I., 2016. Local Content, Trade and Investment: Is There Policy Space Left for Linkages Development in Resource Rich Countries? (No. 205). Discussion Paper.

"Solutions." Briefing No. 5, Local Content Solutions. http://www.localcontentsolutions.com/pdf/SolutionsSeries5.pdf.

Sovacool, B.K., Walter, G., Van de Graaf, T. and Andrews, N., 2016. Energy Governance, Transnational Rules, and the Resource Curse: Exploring the Effectiveness of the Extractive Industries Transparency Initiative (EITI). *World Development*, 83, pp. 179–192.

Sturesson, A. and Zobel, T., 2015. The Extractive Industries Transparency Initiative (EITI) in Uganda: Who Will Take the Lead When the Government Falters? *The Extractive Industries and Society*, 2(1), pp. 33–45.

Taylor, S.D., 2018. Africa and Foreign Direct Investment. In *The Palgrave Handbook of African Politics, Governance and Development* (pp. 709–722). New York: Palgrave Macmillan.

Tordo, S., Warner, M., Manzano, O. F. and Anouti, Y. 2013. *Local Content Policies in the Oil and Gas Sector*. Washington, DC: World Bank Study, World Bank.

8

Regional Content and Beyond

8.1 Introduction

Sub-Saharan African local content regimes are weak, inconsistent, not harmonised with other policies and unclear. One of the biggest hurdles that Sub-Saharan African countries face is the need to replicate the results of local content policies found in developed countries such as Norway and Brazil into their domestic economies.[1] The merge of energy justice into regional content offers innovative solutions to several challenges such as marginalisation, elite capture, corruption, poor infrastructure, lack of capital and expertise within the local population. How can developing countries crack through the enclave and maximise the benefits from the oil and gas resources? To avoid the under development route that is the trajectory of numerous oil and gas countries in Sub-Saharan Africa. After a thorough analysis of local content and regional content regimes. We conclude by discussing the local content implementation gap, the nature of the local content regimes found Sub-Saharan Africa.

[1]Thurber, M.C., Hults, D.R. and Heller, P.R., 2011. Exporting the "Norwegian Model": The Effect of Administrative Design on Oil Sector Performance. *Energy Policy*, 39(9), pp. 5366–5378.

R. S. Muhongo, *Energy Justice*, Energy, Climate and the Environment,
https://doi.org/10.1007/978-3-030-61338-9_8

This analysis facilitates the research to explore what recommendations match the future designs of regional content policy. As much as we focus on the oil and gas industry, the skills obtained and technology transferred from the international oil companies. Must spill-over into other sectors and propel economic development in Sub-Saharan Africa.

The oil and gas industry is integrated into global value chains, and the current oil price shocks of 2020 and COVID-19 pandemic have accelerated the gears of the energy transition. Sub-Saharan Africa has neither the technology nor the expertise to gain form this transition. The shift from fossil fuels to cleaner energy has brought about an increasing demand for certain minerals in Sub-Saharan Africa.[2] Countries such as the Democratic Republic of Congo, Angola, South Africa, Tanzania, Uganda, Kenya and the Central African Republic all have these minerals. A regional content policy would incubate the fast change of technology and industry needs. Since the policy pools resources and gives individual countries absorptive capacity to crack through the enclave of the extractive industry.

8.2 Local Content Regimes in Sub-Saharan Africa

Mature oil and gas producers in Sub-Saharan Africa such as Nigeria and Angola designed, adopted and implemented national content policies. The design of local content policies in the 1980s and the current design of local content policies is entirely different for several emerging African oil producers, namely Uganda, Mozambique, Tanzania, Kenya and Liberia. Their local content policies are based on global markets and the nature of the economy. While for the mature oil and gas producing countries, the policies were more nationalistic. Angola and Nigeria implemented the "Indigenisation policy" that advocated for ownership of the oil and gas resources without consideration of the prevailing economic factors that could help the sector grow. Indigenization in

[2]Heffron, R.J., 2020. The Role of Justice in Developing Critical Minerals. *The Extractive Industries and Society*.

Nigeria has been fought bitterly by foreign capital, and particularly by the Chamber of Commerce and Industry in Lagos.[3] Ultimately, the policies failed to give the Nigerian state significant influence over the industry and were all about elite oil rent capture.[4] After its independence in 1975, Angola nationalised the Portuguese company ANGOL de Lubrificantes e Combustíveis, establishing the Sociedade Nacional de Combustíveis (Sonangol), a national oil business.[5] International companies were permitted to participate in Joint Ventures and Contractor Agreements with Sonangol in Angola, but until 1991 they were only legally allowed to own 49 per cent of each company.[6] Given the relative strength and efficiency of Sonangol, Angola was far more effective in maintaining national control over its oil industry through Sonangol.[7] At the beginning of the 1980s, Angola passed several laws which set targets for Angolans employed by international firms and established a mandatory structure for training and promotion of Angolan employees. However, the IOCs completely ignored specific laws and regulations.[8] The oil industry continued to be an enclave industry. Early attempts at imposing national control represented the dual aims of increasing dominance over a strategic sector and rising developmental benefits.[9] These ambitions were eventually superseded by the self-interest of the elite and the international capital resistance. Although the ruling party in Angola faced a struggle for its very survival and local linkages were undoubtedly out of the question during the civil war, Sonangol arguably could insist

[3]Ovadia, J.S., 2016. Local Content Policies and Petro-Development in Sub-Saharan Africa: A Comparative Analysis. *Resources Policy*, 49, pp. 20–30.

[4]Ovadia, J.S., 2016. Local Content Policies and Petro-Development in Sub-Saharan Africa: A Comparative analysis. *Resources Policy*, 49, pp. 20–30

[5]Ovadia, J.S., 2014. Local Content and Natural Resource Governance: The Cases of Angola and Nigeria. *The Extractive Industries and Society*, 1(2), pp. 137–146.

[6]Ibid.

[7]De Oliveira, R.S., 2007. Business success, Angola-Style: Postcolonial Politics and the Rise and Rise of Sonangol. *Journal of Modern African Studies*, pp. 595–619.

[8]Ovadia, J.S., 2016. Oil-Backed Capitalist Development in the Global South: A Case of Positive Oil Exceptionalism. *Energy, Capitalism and World Order: Towards a New Agenda in International Political Economy*. Palgrave Macmillan.

[9]Ibid.

on more Angolanisation—and certainly more training of Angolans—if it had put long-term interests ahead of short-term payoffs.[10]

Ghana found the "Jubilee field" in 2007; the Ghanaian government passed Ghana's new Local Content Law in 2013 (GLCL). Ghana has missed opportunities because of not placing local content laws and regulations. Opportunities in the mining industry, telecommunications, and the West African gas pipeline and gas infrastructure. The authorities were not going to miss this opportunity again. The government set up the Ghana Petroleum Commission that emphasised on revenue rather than value addition. In Uganda, before the discovery of oil and gas resources. Uganda had no local content policy, no guiding mechanism to provide a deliberate first preference to local Ugandans. Uganda acknowledged its lack of capacity and adopted a flexible national content policy. Mozambique passed a Petroleum Law that advocates for local companies and joint ventures. The Law also calls for the provisions of goods and services for the local population. The law also has provisions for training of Mozambican nationals. Tanzania's first Local Content Policy released in 2014 only "encouraged" international oil companies to work with local Tanzanians. The draft that was enacted by the Petroleum Act of 2015 put the regulations and provisions into law. The local content policy was dual in-nature in that it called for both the community and national content. Kenya developed a Petroleum Bill and Local Content Regulations in 2015 that is still pending. The provisions also call for technology transfer, employment and training, research and development as well as legal and financial services. Oil and gas economies in Sub-Saharan Africa, have implemented local content policies to capture more than oil rents from the oil and gas industry. But the policies have weaknesses that must be addressed, in-comparison to best international practice. The national content policies are not strict on beneficial ownership of local companies leading to fronting of local suppliers sub-contracting to foreign investors. The country fails to apply adequate fiscal policies because of the fronting. The national content policies are also not strict on the import supplier base. Most entrepreneurs from Sub-Saharan Africa do

[10]Ovadia, J.S., 2014. Local Content and Natural Resource Governance: The Cases of Angola and Nigeria. *The Extractive Industries and Society*, 1(2), pp. 137–146.

not manufacture the goods that are supplied to the oil and gas industry. They mostly import these goods from countries like China, Taiwan, Japan and India. Becomes a challenge in building linkages and technology transfer, because there is no value created or added within the economy. The national content policies in Sub-Saharan Africa do not emphasise on any "low-hanging fruits" to guide the local supplier base on what should gain preference in the oil and gas industry. Such challenges make it difficult for the benefits of the industry to trickle down to the marginalised communities. Put together the lack of having an absorptive capacity, and the indigenous capitalist elite has held the oil and gas for ransom. Such elite capture in oil and gas economies has challenged the development of SMEs. We must assess the pitfalls of local content policies against international best practices.

8.3 Lessons from Local Content Best International Practise

8.3.1 Introduction

This section illustrates and explores the different lessons drawn from the cross-country analysis undertaken through a case study methodology in this book. The lessons are obtained from Norway, and Brazil to derive better recommendations and reach a viable conclusion. The lessons derived from Norway, Brazil and Nigeria are provided below.

a. Lessons from Norway

Norway has been the poster boy for good local content policies practice in the world. The success story of Norwegian local content policy is due to various factors that are in the Norwegian economy. Firstly, the Norwegian local content policy adopted a protectionist policy in their oil and gas development.[11] Secondly, Norway was already a technological frontier during the discovery of oil and gas resources.[12] Thirdly Norway was also in a strong commercial position in the marine industry. Fourthly Norway had a political consensus in governing there oil and gas resources.[13] Lastly, Norway had lucky timing since the oil and gas industry had not developed mature technology in offshore oil and gas operations.[14] This gave room for Norway to develop its offshore technology along with the international oil companies. Though the provision did not explicitly define what was meant of local content in Norway, the international oil companies were aware of the policymaker's intentions with regard to local content provisions. Norway separated the oil and gas sector into different segments so as to facilitate efficiency in the sector.[15] Norway separated the roles of the main functionaries of the oil and gas industry. The Ministry of Petroleum and Energy (MPE), the regulatory agency Norwegian Petroleum Directorate and the state oil company Statoil.[16] The Norwegian government also set up an entity responsible for the financial interest of the Government in the oil and gas industry known as SDFI/Petoro.[17] When Norway discovered oil and gas resources, the country had a diversified economy. The population was highly educated with a high standard of living.[18] Hence Norway was in no rush to exploit and develop the oil and gas resources since the country

---

11 Heum, P., 2008. Local Content Development: Experiences from Oil and Gas Activities in Norway.

12 Thurber, M.C., Hults, D.R. and Heller, P.R., 2011. Exporting the "Norwegian Model": The Effect of Administrative Design on Oil Sector Performance. *Energy Policy*, 39(9), pp. 5366–5378.

13 Thurber, M.C., Hults, D.R. and Heller, P.R., 2011. Exporting the "Norwegian Model": The Effect of Administrative Design on Oil Sector Performance. *Energy Policy*, 39(9), pp. 5366–5378.

14 Heum, P., 2008. Local Content Development: Experiences from Oil and Gas Activities in Norway.

15 Al-Kasim, F., 2006. *Managing Petroleum Resources: The "Norwegian model" in a Broad Perspective* (Vol. 30). Oxford Institute for Energy Studies.

16 Al-Kasim, F., 2006. *Managing Petroleum Resources: The "Norwegian Model" in a Broad Perspective* (Vol. 30). Oxford Institute for Energy Studies.

17 Engen, O.A., 2009. *The Development of the Norwegian Petroleum Innovation System: A Historical Overview*, ed. Fagerberg et al. (pp. pp. 179–207). Oxford Handbook of Innovation.

18 Engen, O.A., 2009. *The Development of the Norwegian Petroleum innovation System: A Historical Overview*, ed. Fagerberg et al. (pp.179–207). Oxford Handbook of Innovation.

had socio-economic stability. This gave Norway time to prepare their local capability to integrate the oil and gas industry. Norway adopted an approach that concentrated on the procurement of local goods and services. But only when these goods and services could be competitive in quality, price and delivery date.

Norway also developed a research and development framework between the international oil companies and Norwegian institutions.[19] Local content regulations were implemented under the strict supervision of "watchdog agencies" such as the Goods and Services Office (GSO).[20] This watchdog agency ensured that the international oil companies complied with local content provisions. Since the oil and gas industry is an enclave industry whereby most local companies cannot enter in the oil and gas value chain. Norway unbundled the large contracts into smaller packages to allow the integration of small-medium enterprises from Norway.[21] Norway was a democratic country with political stability. The different political factions in Norway had a consensus on oil and gas policies such as the Ten Oil Commandments, the petroleum fund and the management of Statoil.[22]

b. Lessons from Brazil

Brazil's political approach to governing the oil and gas resources was resource nationalistic. This meant that state participation in the oil and gas industry was predominantly undertaken by Petrobras that had a monopoly for both the downstream and upstream operations in Brazil.[23] Petrobras' dominant position was confronted in 1975, during the former

[19]Ibid.

[20]Noreng, Ø., 2004. Norway: Economic Diversification and the Petroleum Industry. *Middle East Economic Survey*, 47(45).

[21]Thurber, M.C., Hults, D.R. and Heller, P.R., 2011. Exporting the "Norwegian Model": The Effect of Administrative Design on Oil Sector Performance. *Energy Policy*, 39(9), pp. 5366–5378.

[22]Thurber, M.C., Hults, D.R. and Heller, P.R., 2011. Exporting the "Norwegian Model": The Effect of Administrative Design on Oil Sector Performance. *Energy Policy*, 39(9), pp. 5366–5378.

[23]Mendonça, R.W. and de Oliveira, L.G., 2013. Local Content Policy in the Brazilian Oil and Gas Sectoral System of Innovation. *Latin American Business Review*, 14(3–4), pp. 271–287.

general Geisel's period as president, large areas of the Brazilian continental shelf were opened up for international oil companies.[24] Politics and Petrobras have always been aligned together, governing Brazil's oil and gas industry. This led to the Lava Jato (Car Wash) scandal that involved Petrobras.[25] The scandal put an end to Lula da Silva and his Workers' Party (PT) deepening of state capitalism as a development strategy.[26] Local content in Brazil is addressed in the bidding rounds for concession and the conclusions of Production Sharing Contracts. The bidding international oil companies have to comply with a minimum local content requirement to win a bid.[27] Local content requirements are defined and assessed based on percentages. Concessionaires have to ensure a specified percentage of goods and services are sourced within the local market in the execution of their concession and or production sharing contracts. In 1997 Local Content policy was adopted in Brazil with the approval of the Petroleum Law that followed a neoliberal approach for an open market ending the State's monopoly.[28]

In 1999–2004 Brazil's local content regime from Rounds 1 to 6 set the bidders placed unrestricted offers on local content percentage, and the international oil companies would confirm the requirements.[29] In 2005 saw the most significant changes in local content, were in the 7th Bidding Round.[30] There was an introduction of a Local Content Table, featuring items on all aspects of the E&P cycle. Each item was associated with minimum percentages set by the government. Furthermore,

[24] Mendonça, R.W. and de Oliveira, L.G., 2013. Local Content Policy in the Brazilian Oil and Gas Sectoral System of Innovation. *Latin American Business Review*, 14(3–4), pp. 271–287.

[25] Arruda de Almeida, M. and Zagaris, B., 2015. Political Capture in the Petrobus Corruption Scandal: The Sad Tale of an Oil Giant. *Fletcher Forum of World Affairs*, 39, p. 87.

[26] Ibid.

[27] Zacour, C., Zuma Pereira, T., Lima Rocha Cristofaro, A. and Ferreira Francisco, F., 2012. Petrobras and the New Regulatory Framework for the Exploration and Production of Oil and Natural Gas in the Brazilian Pre-Salt Region. *Journal of World Energy Law and Business*, 5(2), pp. 125–138.

[28] Antolín, M.J.P. and Cendrero, J.M.R., 2013. How Important Are National Companies for Oil and Gas Sector Performance? Lessons from the Bolivia and Brazil Case Studies. *Energy Policy*, 61, pp. 707–716.

[29] Filho, N., 2017. Brazil's Oil & Gas Local Content Policy: Lessons Learned. Institute of the Americas.

[30] Ibid.

the inclusion of a Local Content Textbook as a guideline to accurately calculate the given percentages. It also introduced Local Content certification services and a new methodology to conduct these activities[31] from the start of the Bidding Rounds in 1999 till the 2013 Brazilian economic crisis. The local content policy gained substantial results in promoting local companies in the oil and gas sector and facilitated the employability in many sectors within the supply chain.[32]

However, Brazil had an international competitive marine industry. The country later faced industrial failure due to different political motives at the time. The industry was revitalized based on funds granted by government programs and increased demand from Local Content commitments. Due to the given incentives the industry employed over 70,000 by 2013 alone. The Local Content Policy developed other industrial sectors as well despite positive results in terms of increased local participation in all areas of the supply chain. The process has not been smooth, and several decisions were met with resistance from the industry.[33] The policy has been faced with fierce criticism that has led to a policy review. Whereby the oil and gas market participants called for flexible Local Content terms. This is because the stringent local content requirements led to higher costs and hindered further investment that resulted in the minimal attraction of FDI. The changes led to Local Content no longer being a central component of the bidding factor, but the percentage assessment criteria were left in the contractual agreements.[34] Another landmark revision of the local content policy is the introduction of the Program to Encourage Supply Chain Competitiveness and the Development and Improvement of Suppliers in the Oil and

[31] Almada, L.P. and Parente, V., 2013. Oil & Gas Industry in Brazil: A Brief History and Legal Framework. *Panorama of Brazilian Law*, 1(1), pp. 223–252.

[32] Warner, M., 2017. *Local Content in Procurement: Creating Local Jobs and Competitive Domestic Industries in Supply Chains*. Routledge.

[33] Relaxing of Brazil local content rules will boost oil output capacity: report. https://www.reuters.com/article/us-brazil-presalt/relaxing-of-brazil-local-content-rules-will-boost-oil-output-capacity-report-idUSKBN1JG28T.

[34] Relaxing of Brazil local content rules will boost oil output capacity: report. https://www.reuters.com/article/us-brazil-presalt/relaxing-of-brazil-local-content-rules-will-boost-oil-output-capacity-report-idUSKBN1JG28T.

Natural Gas Sector (PEDEFOR).[35] This programme is meant to stimulate the local supply chain and provide incentives for operators and service companies that invest large sums of money towards developing the local Brazilian industry. It should be noted that Local Content must offer alternative methods to stimulate adherence to its requirements, such as capital investments. The current Brazilian Local Content policy only acknowledges local content as the percentage of domestic goods and services rendered to the oil and gas industry. The PEDEFOR program seeks to expand this definition to include investments in new operational plants, R&D centres and investments to bring new technologies to be applied in the Brazilian oil and gas industry.[36]

8.4 Regional Content Policy and Future Considerations

In Sub-Saharan Africa, the proliferation of local content policies has not yielded the results similar to Norway and Brazil. But in this book, it has been established that a regional content policy could provide new avenues that national and community content policies have not yet been able to offer. A regional content policy would replace the strict and hard regulatory policies following a more value-based approach. A regional content policy creates more room for in-country shared value creation. The agglomerations of factors of production in Sub-Saharan Africa regional economic blocs. Create linkages for international institutions and oil and gas economies sharing interest, not excluding everyone else. The most recent oil price shock and COVID-19 pandemic have shown that oil and gas countries can no longer rely on the oil and gas rents to bring about substantial social-economic development. Ovadia states that "only an approach that promotes economic diversification and plans for a future beyond hydrocarbons can serve as a justification for taking those hydrocarbons out of the ground. However, it is precisely

[35]Tordo, S., Warner, M., Manzano, O. and Anouti, Y., 2013. *Local Content Policies in the Oil and Gas Sector*. The World Bank.

[36]Ibid.

when the price of oil falls that oil companies make new efforts to sacrifice local content in the name of the bottom line".[37] The oil price shock also disrupted global value chains due to the lockdown measures adopted in different countries. Having a regional content policy would strengthen regional value chains making the region more resilient from uncertain global shocks. Does the main question remain of how regional content policies can crack through the enclave of the oil and gas industry? We conclude by applying the tenets of energy justice in the adopting of these policies for the sole purpose of cracking the enclave.

While cracking through the enclave of the oil and gas industry, one must also consider why energy justice is needed in designing and implementing a regional content policy. The oil and gas industry in Sub-Saharan Africa has so far created the indigenous capitalist elite class in countries like Nigeria, Angola, Equatorial Guinea and Ghana. The resource curse epidemic has been going on for years in Sub-Saharan Africa because of inefficient repressive policies. In reality, though the natural resources belong to the whole population, only a few people and group of the population enjoy the fruits of these resources in Sub-Saharan Africa. Hence, there is no justice in the distribution of resources, the recognition of marginalised communities and the implementation of natural resource governance policies such as local content policy. The key reason to have justice, therefore, is to ensure provisions of fairness and equality in society,[38] with the sole purpose of resolving the growing inequality in Sub-Saharan Africa. Considering that local Sub-Saharan population has missed out on opportunities that were brought by the mining sector, the oil and gas sector is another industry that local populations are failing to engage in due to its enclave nature. The scare comes with the energy transition for Sub-Saharan Africa, as we move away from fossil fuels to cleaner sources of energy that brings the rise of demands of critical minerals. It is clear that for there to be a just transition. We must adopt a regional content policy to incubate the skills, pool the resources for quick indigenous integration in the oil and gas

[37]Ovadia, J.S., 2016. Local Content Policies and Petro-Development in Sub-Saharan Africa: A Comparative Analysis. *Resources Policy*, 49, pp. 20–30.

[38]Heffron, R.J., 2020. The Role of Justice in Developing Critical Minerals. *The Extractive Industries and Society*.

value chain before we lose another opportunity as we lost the Gold Rush opportunity of Southern Africa. Considering the massive movement and role Sub-Saharan resources in the development of the low-carbon economy with the majority of the countries have signed up the Paris COP21 Agreement.[39] Now more than ever Sub-Saharan countries must develop regional content policies to meet the demands of this withering oil and gas industry. We undertake the Just Framework that researchers have previously applied to the determine practical and interdisciplinary perspective and ensure the researcher assesses the justices issues specific to the problem, wider societal justice issues in terms of universal justice, the geography issues such as where we need to consider global supply chains such as in the oil and gas industry. The just framework covers five key elements of justice; we adopt these elements into the regional content policy and oil and gas lifecycle. We must look at the different tenets and their applicability in the regional content policy. Procedural justice—the focus here is on the legal process and the steps to create a regional content policy. Recognition justice—are rights recognized for different groups in society and the marginalised communities that are host communities to these oil and gas resources Cosmopolitanism justice—this stems from the belief we are all citizens of the world; the oil and gas industry is an industry with global and regional value chains. That is important in regional content development. Restorative justice—any injustice caused by the oil and gas industry should be rectified, and it focuses on the need for enforcement of particular las. Local content policies are not restorative. A regional content policy includes more accountability due to increased transparency with regional stakeholders and regional watchdogs.

8.5 Conclusion

The extractive industry in Sub-Saharan Africa has been pivotal in global development. But Sub-Saharan Africa has little to show for it. From the South African Gold Rush of the 1800s, the oil and gas discoveries of the

[39] Ibid.

1940s and 1950s. It is clear that the common African is yet to receive his/her fair share of the pie. Many theorists and researchers have developed "excuses" as to why Sub-Saharan Africa is not developing though the region has vast natural resources. Poor natural resource management, democracy, corruption, lack of infrastructure, lack of capital, minimal linkages, lack of political will, neo-colonial tactics, lack of a skilled workforce. All these are some of the reasons to name a few. But in our interdisciplinary research, we find that we lack justice in our legal and regulatory resource management mechanisms. Governments have devised means to intervene in the oil and gas sector for decades. Adopting different production development policies that can materialise the probable benefits of the extractive industry into a vehicle of economic development for any community. This led to the proliferation of local content regimes from the 1970s up to date. Mature oil and gas countries like Nigeria and Angola first adopted the indigenisation policies that gave ownership of the resources to the local citizens. But there was no due process, no recognition of marginalised communities and no equitable distribution of concessions. This led to under-development, civil unrest, the capture of the political elite and most of all, the development of redundant policies.

Sub-Saharan Africa oil and gas producers have failed to emulate the local content poster boys of Norway and Brazil because we do not have the same technical and technological advancement that can incubate all the R&D in our local economies. As Norway and Brazil concentrated on R&D, developing and creating a monopoly through the national oil company, minimising taxes to attract foreign investments, have political stability with regards to the industry, develop on the skills that the local workforce could obtain and most of all emphasised on value addition. Sub-Saharan countries just concentrated on the oil rents from the oil and gas companies. The oil and gas industry in Sub-Saharan Africa was built without any foundation. Norway paced their exploitation because they have a robust fishing and shipping industry. But in Sub-Saharan Africa, we exhaust and deplete our reasons quickly to bankroll the government coffers. Hence, no tangible outcome has been seen or felt due to our oil and gas resources. This book calls for regionalisation of local content policies. The creation of regional content policy, because Sub-Saharan countries do not have the same characteristics and capability as Norway and Brazil. The agglomeration and pooling of skills, resources and will. Will facilitates larger spill-overs and linkages as well as faster developmental infrastructure with huge trickle-down effects. Regional bodies like ECOWAS, SADC, EAC, COMESA, and the AU. Should join forces to achieve justice in the Sub-Saharan oil and gas life cycle. The competing development notions of industrialisation but in a low-carbon economy, the shift from fossil fuels to cleaner sources of energy, the changes in technology ad key industries that has led to demands in critical minerals, the Post-COVID-19 developmental and industrial tactics that have disrupted global and regional value chains. All these developments and competing interest cannot be met by individual Sub-Saharan African countries. There needs to be a pooling of resources, skill and capital through a regional content policy in a just and fair way.

References

Al-Kasim, F., 2006. Managing Petroleum Resources: The "Norwegian Model" in a Broad Perspective (Vol. 30). Oxford Institute for Energy Studies.

Almada, L.P. and Parente, V., 2013. Oil & Gas industry in Brazil: A Brief History and Legal Framework. *Panorama of Brazilian Law*, 1(1), pp. 223–252.

Antolín, M.J.P. and Cendrero, J.M.R., 2013. How Important Are National Companies for Oil and Gas Sector Performance? Lessons from the Bolivia and Brazil Case Studies. *Energy Policy*, 61, pp. 707–716.

Arruda de Almeida, M. and Zagaris, B., 2015. Political Capture in the Petrobus Corruption Scandal: The Sad Tale of an Oil Giant. *Fletcher Forum of World Affairs*, 39, p. 87.

De Oliveira, R.S., 2007. Business Success, Angola-Style: Postcolonial Politics and the Rise and Rise of Sonangol. *Journal of Modern African Studies*, pp. 595–619.

Filho, N., 2017. *Brazil's Oil & Gas Local Content Policy: Lessons Learned.* Institute of the Americas

Engen, O.A., 2009. *The Development of the Norwegian Petroleum Innovation System: A Historical Overview*, ed. Fagerberg et al. (pp. 179–207). Oxford Handbook of Innovation.

Heffron, R.J., 2020. The Role of Justice in Developing Critical Minerals. *The Extractive Industries and Society*.

Heum, P., 2008. Local Content Development: Experiences from Oil and Gas Activities in Norway.

Mendonça, R.W. and de Oliveira, L.G., 2013. Local Content Policy in the Brazilian Oil and Gas Sectoral System of Innovation. *Latin American Business Review*, 14(3–4), pp. 271–287.

Noreng, Ø., 2004. Norway: Economic Diversification and the Petroleum Industry. *Middle East Economic Survey*, 47(45).

Ovadia, J.S., 2014. Local Content and Natural Resource Governance: The Cases of Angola and Nigeria. *The Extractive Industries and Society*, 1(2), pp. 137–146.

Ovadia, J.S., 2016. Local Content Policies and Petro-Development in Sub-Saharan Africa: A Comparative Analysis. *Resources Policy*, 49, pp. 20–30.

Ovadia, J.S., 2016. Oil-Backed Capitalist Development in the Global South: A Case of Positive Oil Exceptionalism. *Energy, Capitalism and World Order: Towards a New Agenda in International Political Economy*. Palgrave Macmillan.

Relaxing of Brazil local content rules will boost oil output capacity: report. https://www.reuters.com/article/us-brazil-presalt/relaxing-of-brazil-local-content-rules-will-boost-oil-output-capacity-report-idUSKBN1JG28T.

Thurber, M.C., Hults, D.R. and Heller, P.R., 2011. Exporting the "Norwegian Model": The Effect of Administrative Design on Oil Sector Performance. *Energy Policy*, 39(9), pp. 5366–5378.

Tordo, S., Warner, M., Manzano, O. and Anouti, Y., 2013. *Local Content Policies in the Oil and Gas Sector*. The World Bank.

Warner, M., 2017. *Local Content in Procurement: Creating Local Jobs and Competitive Domestic Industries in Supply Chains*. Routledge.

Zacour, C., Zuma Pereira, T., Lima Rocha Cristofaro, A. and Ferreira Francisco, F., 2012. Petrobras and the New Regulatory Framework for the Exploration and Production of Oil and Natural Gas in the Brazilian Pre-Salt Region. *Journal of World Energy Law and Business*, 5(2), pp. 125–138.

Bibliography

A Guide to Getting Started in Local Procurement: For Companies Seeking the Benefits of Linkages with Local SMEs. International Financial Corporation. http://www.ifc.org/wps/wcm/connect/03e40880488553ccb09cf26a6515bb18/IFC_LPPGuide_PDF20110708.pdf?MOD=AJPERES.

Abidin, M.Z., 2001. Competitive Industrialization with Natural Resource Abundance: Malaysia. In R.M. Auty (Ed.), *Resource Abundance and Economic Development* (pp. 147–164). Oxford University Press.

Ablo, AD., 2019. Actors, Networks and Assemblages: Local Content, Corruption and the Politics of SME's Participation in Ghana's Oil and Gas Industry. International Development Planning Review.

Ablo, A.D., 2015. Local Content and Participation in Ghana's Oil and Gas Industry: Can Enterprise Development Make a Difference? *The Extractive Industries and Society*, 2(2).

Ackah-Baidoo, A., 2012. Enclave Development and 'Offshore Corporate Social Responsibility': Implications for Oil-Rich Sub Saharan Africa. *Resources Policy*, 37(2).

ADB (African Development Bank), 2013. African Economic Outlook: Structural Transformation and Extractives. African Development Bank, the Development Centre of the Organisation for Economic Co-operation and Development and United Nations.

R. S. Muhongo, *Energy Justice*, Energy, Climate and the Environment, https://doi.org/10.1007/978-3-030-61338-9

ADB (African Development Bank), 2017. African Economic Outlook: Structural Transformation and Extractives. African Development Bank, the Development Centre of the Organisation for Economic Co-operation and Development and United Nations.

Adekalu, S.O., Oludeyi, O.S., Genty, K.I. and Wolo, A., 2013. Petroleum Technology Development Fund (PTDF) Mandates and Human Capacity Development in Nigeria: Benefits for Nigerian Youths. *International Journal of Research in Management Issue*, 3.

Adetayo, S., 2020. The Ethics of State Capture: Dangote and the Nigerian State. In *The Palgrave Handbook of African Social Ethics* (pp. 371–388). Cham: Palgrave Macmillan.

Agreement on the Exploration, Development and Production Sharing for the Shah Deniz Prospective Area in the Azerbaijan Sector of the Caspian Sea, 1996. http://www.bp.com/liveassets/bp_internet/bp_caspian/bp_caspian_en/STAGING/local_assets/downloads_pdfs/pq/SD_PSA.pdf.

Aiginger, K. and Firgo, M., 2015. Regional Competitiveness Under New Perspectives.

Akinyemi, O., Efobi, U., Osabuohien, E. and Alege, P., 2019. Regional Integration and Energy Sustainability in Africa: Exploring the Challenges and Prospects for ECOWAS. *African Development Review*, 31(4), pp. 517–528.

Al Chukwuma Okoli. 2019. Oil Pipeline Vandalism in the Niger Delta: Need, Greed and Grievance Factors.

Alesina, A., et al., 2002. Fractionalization. Harvard Institute of Economic Research Discussion Paper No 1959, p. 8, online: http://dash.harvard.edu/bitstream/handle/1/4553003/alesinassrn_fractionalization.pdf?sequence=2.

Al-Kasim, F., 2006. *Managing Petroleum Resources: The "Norwegian Model" in a Broad Perspective* (Vol. 30). Oxford Institute for Energy Studies.

Almada, L.P. and Parente, V., 2013. Oil & Gas Industry in Brazil: A Brief History and Legal Framework. *Panorama of Brazilian Law*, 1(1), pp. 223–252.

Alzina, Virginia, et al. [2004]. Enclave Projects? What Are They and What Can Be Done? Document of the Inter-American Development Bank.

Ambrose, K.W., 2010. Democracy and Development in Uganda. *OIDA International Journal of Sustainable Development*, 2(4), pp. 95–104.

Amendolaigne, V., Boly, A., Coniglio, N. D., and Prota, F., 2013. FDI and Local Linkages in Developing Countries: Evidence from Sub-Saharan Africa. *World Development*, 50, 41–56.

Andersen, S.S., 1993. *The Struggle Over North Sea Oil and Gas: Government Strategies in Denmark, Britain, and Norway*. A Scandinavian University Press Publication.

Anderson, D.M. and Browne, A.J., 2011. The Politics of Oil in Eastern Africa. *Journal of Eastern African Studies*, 5(2), pp. 369–410.

Antolín, M.J.P. and Cendrero, J.M.R., 2013. How Important Are National Companies for Oil and Gas Sector Performance? Lessons from the Bolivia and Brazil Case Studies. *Energy Policy*, 61, pp. 707–716.

Araujo, F.C. and Leoneti, A.B., 2019. How Attractive Is Brazil's Oil and Gas Regulatory Framework to Investors? The Extractive Industries and Society.

Arruda de Almeida, M. and Zagaris, B., 2015. Political Capture in the Petrobus Corruption Scandal: The Sad Tale of an Oil Giant. *Fletcher* Forum of *World* Affairs, 39, p. 87.

Asiago, B.C., 2016. Fact or Fiction: Harmonising and Unifying Legal Principles of Local Content Requirements. *Journal of Energy & Natural Resources Law*, 34(3), pp. 337–360.

Asiago, B.C., 2017. Norwegian Local Content Model a Viable Solution. *US-China Law Review*, 14, p. 471.

Bahoo, S., Alon, I. and Paltrinieri, A., 2019. Sovereign Wealth Funds: Past, Present and Future. *International Review of Financial Analysis*, p. 101418.

Balouga, J., 2012. Nigerian Local Content: Challenges and Prospects. *International Association for Energy Economics*, 4, pp. 23–26.

Barczewski, B., 2013. How Well Do Environmental Regulations Work in Kenya?: A Case Study of the Thika Highway Improvement Project. Center for Sustainable Urban Development.

Basnett, Y., 2013. Labour Mobility in East Africa: An Analysis of the East African Community's Common Market and the Free Movement of Workers. *Development Policy Review*, 31(2), pp. 131–148.

Boldbaatar, D., Kunz, N.C. and Werker, E., 2019. Improved Resource Governance Through Transparency: Evidence from Mongolia. The Extractive Industries and Society.

Booth, D., Cammack, D., Kibua, T., Kweka, J. and Rudaheranwa, N., 2007. East African Integration: How Can It Contribute to East African Development? Overseas Development Institute Project Briefing.

Booth, D., et al., East African Prospects: An Update on the Political Economy of Kenya, Rwanda, Tanzania and Uganda.

Brazil Model Contract 2015, Art.1.3.15.

Briffault, R., 1999. Localism and Regionalism, Columbia Law School Public Law and Legal Theory Working Paper, No 1, August, 11, p. 18.

Briffault, R., 2000. Localism and Regionalism. *Buffalo Law Review*, 48, p. 1.
Brooks, S.M. and Kurtz, M.J., 2016. Oil and Democracy: Endogenous Natural Resources and the Political "Resource Curse". *International Organization*, 70(2), pp. 279–311.
Budd, L. and Hirmis, A., 2004. Conceptual Framework for Regional Competitiveness. *Regional Studies*, 38(9), pp. 1015–1028.
Bullard, R., 2000. Environmental Justice in the 21st Century. People of Colour Environmental Groups. Directory, pp. 1–21.
Bulmer, S.J., 1998. New Institutionalism and the Governance of the Single European Market. *Journal of European Public Policy*, 5(3), pp. 365–386.
Calignano, G. and Vaaland, T.I., 2018. Local Content in Tanzania: Are Local Suppliers Motivated to Improve? *The Extractive Industries and Society*, 5(1), pp. 104–113.
Campos, J.E. and Pradhan, S. (Eds.), 2007. *The Many Faces of Corruption: Tracking Vulnerabilities at the Sector Level*. World Bank Publications.
Cappelen, Å. and Mjøset, L., 2009. Can Norway Be a Role Model for Natural Resource Abundant Countries? *Chp*, 3, pp. 44–72.
Carroll, A.B., 1979. A Three-Dimensional Conceptual Model of Corporate Performance. *Academy of Management Review*, 4(4), pp. 497–505.
Cavalheiro, G.M. do C., Joia, L.A., and Gonzalves, A.C., 2014. Strategic Patenting in the Upstream Oil and Gas Industry: Assessing the Impact of the Pre-Salt Discovery on Patent Applications in Brazil. *World Patent Information*, 39, pp. 58–68.
Chingarande, A., Mzumara, M. and Karambakuwa, R., 2013. Comparative Advantage and Economic Performance of East African Community (EAC) Member States. *Journal of Economics*, 4(1), pp. 39–46.
Cimino, C., Hufbauer, G.C. and Schott, J.J., 2014, February. A Proposed Code to Discipline Local Content Requirements. Peterson Institute for International Economics Policy Brief No PB14–6, p. 1.
Collier, P., and Venebles, A., 2011. *Plundered Nations? Successes and Failures in Natural Resource Extraction*. Palgrave Macmillan.
Competitive Advantage May Be Defined as: The Ability to Implement a Competitive Strategy so as to Secure a Favourable Position. Clemons, E.K., 2019. Resources, Platforms, and Sustainable Competitive Advantage: How to Win and Keep on Winning. In *New Patterns of Power and Profit* (pp. 93–104). Cham: Palgrave Macmillan.
Corden, W.M., 1985. *Protection, Growth and Trade: Essays in International Economics*. Oxford: Blackwell.

Cruz, A.S. and Mafambissa, F.J., 2016. Industries Without Smokestacks, p. 232.

De Almeida, E. and Martinez-Prieto, D., 2015. The Impact and Effectiveness of Local Content Policy on Oil Exploration and Production in Brazil. In *Oxford Energy Forum* (Vol. 98, No. November 2014).

De Oliveira, R.S., 2007. Business Success, Angola-Style: Postcolonial Politics and the Rise and Rise of Sonangol. *Journal of Modern African Studies*, pp. 595–619.

Doingbusiness.org, 2019. [online] Available at: https://www.doingbusiness.org/content/dam/doingBusiness/media/Annual-Reports/English/DB2019-report_web-version.pdf.

East Africa Economic Outlook, 2018. Macroeconomic Developments Manufacturing's Comparative Advantage and Competitiveness.

Economic Transformation Program 2010 ("ETP"). The ETP Embodies the Government Project to Transform Malaysia into a High-Income Country by 2020. Malaysia Aims to Become a Regional Hub by 2017 and Private Sector Plays a Crucial Role.

EITI, The EITI Standard, 2016. Available at: https://eiti.org/document/standard.

Emmanuel, R., 2016. Optimal Local Content Requirement Policies for Extractive Industries. *Resources Policy*, 50, pp. 244–252.

Employment of Foreign Citizens Non-Resident, Decree 5/95, Ministry of the Interior, Republic of Angola. http://www.sme.ao/attachments/article/209/Decree%20No.%20595%20of%20April%207%20%20Employment%20of%20Foreign%20Citizens%20Non-resident.pdf.

Engen, O.A., 2009. The Development of the Norwegian Petroleum Innovation System: A Historical Overview. Innovation, Path Dependency and Policy: The Norwegian Case, pp. 179–207.

Estevez, A.M., et al. Local Content Initiatives: Enhancing the Subnational Benefits of the Oil, Gas and Mining Sectors.

Evans, M., 1986. The Front-line States South Africa and Southern African Security: Military Prospects and Perspectives.

Farooki, M. and Kaplinsky, R., 2014. Promoting Diversification in Resource-Rich Economies. *Mineral Economics*, 27(2–3), pp. 103–113.

Fayomi, O. and Adeola, G.L., 2016. ECOWAS and Sub-Regional Integration in West Africa: An Appraisal. *Covenant University Journal of Politics and International Affairs*, 3(2).

Ferguson, J., 2005. Seeing Like an Oil Company: Space, Security, and Global Capital in Neoliberal Africa. *American Anthropologist*, 107(3), 377–382.

Filho, N.N. Brazil's Oil & Gas Local Content Policy: Lessons Learned. Institute of The Americas.

Fraser, N., 2009. Social Justice in the Age of Identity Politics. Geographic Thought: A Praxis Perspective, pp. 72–91.

Frederiksen, T., 2019. Political Settlements, the Mining Industry and Corporate Social Responsibility in Developing Countries. *The Extractive Industries and Society*, 6(1), pp. 162–170.

Fuller, S. and McCauley, D., 2016. Framing Energy Justice: Perspectives from Activism and Advocacy. *Energy Research & Social Science*, 11, pp. 1–8.

Gastorn, K., & Masinde, W., 2017. The EAC Common Market. In Ugirashebuja E., Ruhangisa J., Ottervanger T., & Cuyvers A. (Eds.), East African Community Law: Institutional, Substantive and Comparative EU Aspects (pp. 285–292). LEIDEN; BOSTON: Brill. Retrieved from http://www.jstor.org/stable/10.1163/j.ctt1w76vj2.21.

Gelb, A., 2010. Economic Diversification in Resource Rich Countries. Centre for Global Development, pp. 1–23.

Gereffi, G., Humphrey, J. and Sturgeon, T., 2005. The Governance of Global Value Chains. *Review of International Political Economy*, 12(1), pp. 78–104.

Gidado, M.M., 1999. Petroleum Development Contracts with Multinational Oil Firms: The Nigerian Experience. Ed-Linform Services.

GIZ, 2016. Space for Local Content Policies and Strategies. A Crucial Time to Revisit an Old Debate. Report prepared by the Columbia Centre for Sustainable Investment.

Graham, E. and Ovadia, J.S., 2019. Oil Exploration and Production in Sub-Saharan Africa, 1990–Present: Trends and Developments. The Extractive Industries and Society.

Grant, R.M., 1991. Porter's 'Competitive Advantage of Nations': An Assessment. *Strategic Management Journal*, 12(7), pp. 535–548.

Grossman, G.M., 1981. The Theory of Domestic Content Protection and Content Preference. *The Quarterly Journal of Economics*, 96(4), pp. 583–603.

Gustafsson, M.T. and Scurrah, M., 2019. Unpacking the Extractivist State: The Role of Weak State Agencies in Promoting Institutional Change in Peru. *The Extractive Industries and Society*, 6(1), pp. 206–214.

Haas, E.B., 1980. Why Collaborate? Issue-Linkage and International Regimes. *World Politics*, 32(3), pp. 357–405.

Haas, E.B., 1964. *Beyond the Nation-State*. Stanford: Stanford University Press.

Hansen, M. W., and Schaumburg-Müller, H. (Eds.), 2006. *Transnational Corporations and Local Firms in Developing Countries Linking and Upgrading*. CBS Press.

Hansen, M.W., 2014. From Enclave to Linkage Economies? A Review of the Literature on Linkages Between Extractive Multinational Corporations and Local Industry in Africa, DIIS Working Paper, No. 2, Copenhagen, Danish Institute for International Studies.

Hansen, M.W., 2013. Reaping the Rewards of Foreign Direct Investment: Linkages between Extractive Multinational Corporations and Local Firms in Tanzania.

Hansen, M.W., Buur, L., Mette Kjær, A. and Therkildsen, O., 2016, May. The Economics and Politics of Local Content in African Extractives: Lessons from Tanzania, Uganda and Mozambique. In *Forum for Development Studies*, 43(2), pp. 201–228). Routledge.

Hausmann, B.M., Moreira, M. and Rodrik, D., 2006. Coordination Failures, Clusters, and Microeconomic Interventions. Andrés Velasco, p. 1.

Hausmann, R. and Rodrik, D., 2006. *Doomed to Choose: Industrial Policy as Predicament*. John F. Kennedy School of Government, Harvard University, 9.

Hausmann, R., Klinger, B. and Lawrence, R., 2008. *Examining Beneficiation*. Harvard University.

Hazlewood, A., 1979. The End of the East African Community: What Are the Lessons for Regional Integration Schemes? *JCMS: Journal of Common Market Studies*, 18(1), pp. 40–58.

Heffron, R.J., 2014. *Energy Law: An Introduction*. Springer.

Heffron, R.J., 2020. The Role of Justice in Developing Critical Minerals. The Extractive Industries and Society.

Heffron, R.J. and McCauley, D., 2017. The Concept of Energy Justice Across the Disciplines. *Energy Policy*, 105, pp. 658–667.

Heffron, R.J. and Talus, K., 2016. The Development of Energy Law in the 21st Century: A Paradigm Shift? *The Journal of World Energy Law & Business*, 9(3), pp. 189–202.

Heffron, R.J., McCauley, D. and Sovacool, B.K., 2015. Resolving Society's Energy Trilemma Through the Energy Justice Metric. *Energy Policy*, 87, pp. 168–176.

Heffron, R.J., Rønne, A., Tomain, J.P., Bradbrook, A. and Talus, K., 2018. A Treatise for Energy Law. *The Journal of World Energy Law & Business*, 11(1), pp. 34–48.

Hemingway, J., 2019. *The East African Union: Regional integration in the East African Community* (Master's book).

Hettne, B. and Söderbaum, F., 1998. The New Regionalism Approach. *Politeia*, 17(3), pp. 6–21.

Heum, P., 2008. Local Content Development: Experiences from Oil and Gas Activities in Norway. https://openaccess.nhh.no/nhh-xmlui/bitstream/handle/11250/166156/A02_08.pdf?sequence=1&isAllowed=y.

High Expectations as Rwanda and Burundi join East African Community. https://www.voanews.com/a/a-13-2007-07-12-voa37/406029.html.

Hilson, G., 2012. Corporate Social Responsibility in the Extractive Industries: Experiences from Developing Countries. *Resources Policy*, 37(2), pp. 131–137.

Hilson, G., Maconachie, R., 2009. Good Governance and the Extractive Industries in Sub-Saharan Africa. *Mineral Processing and Extractive Metallurgy Review*, 30(1), 52–100.

Hirschman, A.O., 1958. *The Strategy of Economic Development*. Yale University Press.

Hoekman, B. and Martin, W., 2012. Reducing Distortions in International Commodity Markets: An Agenda for Multilateral Cooperation. World Bank Policy Research Working Paper Series. http://www.ipieca.org/resources/good-practice/local-content-strategy-a-guidance-document-for-the-oil-and-gas-industry-1st-edition/. Local content strategy: a guidance document for the oil and gas industry (1st edition).

http://www.tpdc-tz.com/MPSA%20_2008.pdf. Art. 18 of the PSA That Calls for Maximization of Local Goods and Services Where Available on a Competitive Basis.

Huenteler, J., Niebuhr, C. and Schmidt, T.S., 2016. The Effect of Local and Global Learning on the Cost of Renewable Energy in Developing Countries. *Journal of Cleaner Production*, 128, pp. 6–21.

Hufbauer, G.C., Schott, J.J. and Cimino-Isaacs, C., 2013. *Local Content Requirements: A Global Problem* (Vol. 102). Columbia University Press.

Humphreys, M., et al., 2007. Introduction. In *Escaping the Resource Curse*. Columbia University Press.

Humphreys, M., Sachs, J.D., and Stiglitz, J.E. (Eds.), 2007. *Escaping the Resource Curse*. New York: Columbia University Press.

Hunter, T., 2010. *Legal Regulatory Framework for the Sustainable Extraction of Australian Offshore Petroleum Resources. A Critical Functional Analysis*. The University of Bergen.

Hunter, T., 2014. Law and Policy Frameworks for Local Content in the Development of Petroleum Resources: Norwegian and Australian Perspectives on Cross-Sectoral Linkages and Economic Diversification. *Mineral Economics*, 27, p. 115.

Ihua, U.B., Ajayi, C. and Eloji, K.N., 2009, May. Nigerian Content Policy in the Oil and Gas Industry: Implications for Small and Medium-Sized Oil Services Companies. In Proceedings of the 10th Annual Conference, IAABD.

Intergovernmental Forum on Mining, Minerals, Metals and Sustainable Development, 2015. Norway: Horizontal Linkages. Local Context as the Main Driver for Horizontal Linkages.

International Finance Corporation (IFC), 2015. The Art and Science of Benefit Sharing in the Natural Resource.

IPIECA (International Petroleum Industry Environmental and Conservation Association).

IPIECA, 2011. Local Content Strategy: A Guidance Document for the Oil and Gas Industry. Prepared by the International Petroleum Industry Environmental and Conservation Association.

Jackson, J.H., 2000. The Perils of Globalization and the World Trading System. *Fordham Int'l LJ*, 24, p. 371.

Jacob, T. and Pedersen, R.H., 2018. New Resource Nationalism? Continuity and Change in Tanzania's Extractive Industries. *The Extractive Industries and Society*, 5(2), pp. 287–292.

Jenkins, K., McCauley, D. and Forman, A., 2017. Energy Justice: A Policy Approach. *Energy Policy*, 105, pp. 631–634.

Jenkins, K., McCauley, D., Heffron, R., Stephan, H. and Rehner, R., 2016. Energy Justice: A Conceptual Review. *Energy Research & Social Science*, 11, pp. 174–182.

Jesus, F. D., Silva, J.F. and Roitman, T., 2017. Industrial Policy and Local Content: Brazil's Pre-Salt Area Case. Offshore Technology Conference.

Johannes, E.M., Zulu, L.C. and Kalipeni, E., 2015. Oil Discovery in Turkana County, Kenya: A Source of Conflict or Development? *African Geographical Review*, 34(2), pp. 142–164.

Johnson, L. and J. Coleman. 2016. International Investment Law and the Extractive Industries Sector. Briefing Note. Columbia Centre on Sustainable Investment.

Kabananiye, A., 2011. Regional Integration and Sustainable Growth in Sub-Sahara Africa: A Case Study of the East Africa Community (Doctoral dissertation, Northumbria University).

Kaplinsky, R., 2011. Commodities for Industrial Development: Making Linkages Work. Working Paper 01/2011, United Nations Industrial Development Organization.

Kazzazi, A. and Nourri, B., 2012. A Conceptual Model for Local Content Development in Petroleum Industry. *Management Science Letters*, 2(2012), pp. 2165–2174.

Kenya Says Crude Oil Capacity Insufficient for Refinery. https://www.theeastafrican.co.ke/business/Kenya-says-crude-oil-capacity-insufficient-for-refinery/2560-4989226-q3uar1/index.html.

Kessler, A.S., Hansen, N.A. and Lessmann, C., 2011. Interregional Redistribution and Mobility in Federations: A Positive Approach, *Review of Economic Studies*, 78, p. 1345.

Kitson, M., Martin, R. and Tyler, P., 2004. Regional Competitiveness: An Elusive Yet Key Concept? *Regional Studies*, 38(9), pp. 991–999.

Kjær, A.M. and Katusiimeh, M., 2012. Growing But Not Transforming: Fragmented Ruling Coalitions and Economic Developments in Uganda (No. 2012: 07). DIIS Working Paper.

Klueh, U.H., Pastor, G., Segura, A., 2009. Policies to Improve the Local Impact from Hydrocarbon Extraction: Observations on West Africa and Possible Lessons for Central Asia. *Energy Policy*, 37(3), pp. 1128–1144.

Kolstad, I. and Kinyondo, A., 2017. Alternatives to Local Content Requirements in Resource-Rich Countries. *Oxford Development Studies*, 45(4), pp. 409–423.

Kramon, E. and Posner, D.N., 2011. Kenya's New Constitution. *Journal of Democracy*, 22(2), pp. 89–103.

LaBelle, M.C., 2017. In Pursuit of Energy Justice. *Energy Policy*, 107, pp. 615–620.

Lalbahadur, A., A Regional Approach to National Development Priorities in Mozambique? 20 March 2013, South African Institute of International Affairs, online: https://saiia.org.za/research/a-regional-approach-to-national-development-priorities-in-mozambique/.

Lall, S., 2013. Reinventing Industrial Strategy: The Role of Government Policy in Building Industrial Competitiveness. *Annals of Economics & Finance*, 14(2).

Lange, S. and Kinyondo, A., 2016. Resource Nationalism and Local Content in Tanzania: Experiences from Mining and Consequences for the Petroleum Sector. *The Extractive Industries and Society*, 3(4), pp. 1095–1104.

Law 10/79 Mandating Sonangol to Employ and Train Nationals.

Law 291-IV/2010 on Subsoil and Subsoil Users.

Law on Subsoil and Subsoil Users in Kazakhstan, No. 291-IV, June 2010, Republic of Kazakhstan.

Lee, B. and Dupuy, K., 2018. Understanding the Lie of the Land: An Institutional Analysis of Petro-Governance in Tanzania. *Journal of Energy & Natural Resources Law*, 36(1), pp. 85–101.

Lima de Oliveira, R., 2016. Resource-Led Industrial Development in the Oil & Gas Supply Chain: The Case of Brazil.

Linkages to the Resource Sector. The Role of Companies, Government and International Development Cooperation. The Giz. http://ccsi.columbia.edu/files/2016/07/Linkages-to-the-resource-sector-GIZ-CCSI-2016.pdf.pdf.

Littlechild, S. and Vaidya, K., 2019. *Energy Strategies for the UK*. Routledge.

Lloyd, P.J., 1973. *Non-tariff Distortions of Australian Trade*. Canberra: Australian National University Press.

Local Content Bill of Kenya, 2018.

Local Content: Goals and Brief History. Rio de Janeiro: ANP. http://www.anp.gov.br/brasil-rounds/round10/ingles/conteudo_local_Obj_e.asp.

Malaysia on Target to Become Regional Oil and Gas Hub: MPRC. https://www.nst.com.my/business/2017/09/278457/malaysia-target-become-regional-oil-and-gas-hub-mprc.

Marcel, V., Tissot, R., Paul, A. and Omonbude, E., 2016. *A Local Content Decision Tree for Emerging Producers*. Chatham House, The Royal Institute of International Affairs: London, UK.

Maxon, R., 2009. *East Africa: An Introductory History*. West Virginia University Press.

Mbayi, L., 2011. Linkages in Botswana's Diamond Cutting and Polishing Industry. MMCP Discussion Paper. The Open University and University of Cape Town.

McCauley, D., Ramasar, V., Heffron, R.J., Sovacool, B.K., Mebratu, D. and Mundaca, L., 2019. Energy Justice in the Transition to Low Carbon Energy Systems: Exploring Key Themes in Interdisciplinary Research.

McCauley, D.A., Heffron, R.J., Stephan, H. and Jenkins, K., 2013. Advancing Energy Justice: the Triumvirate of Tenets. *International Energy Law Review*, 32(3), pp. 107–110.

McKinsey Global Institute, 2011. *Lions on the Move: The Progress and Potential of African Economies*. McKinsey.

McPherson, C., 2010. 9 State Participation in the Natural Resource Sectors. The Taxation of Petroleum and Minerals: Principles, Problems and Practice, pp. 263–288.

McSherry, B., 2006. The Political Economy of Oil in Equatorial Guinea. *African Studies Quarterly*, 8(3).

Meet Malaysia: Investment Opportunities in Asia's Oil and Gas Hub Malaysia's Mature But Rapidly Growing Oil and Gas Sector Has Turned It into a Regional O&G Hub. http://www.mida.gov.my/home/42/pages/.

Mega Projects Minting Millionaires in Villages. https://www.standardmedia.co.ke/business/article/2001313841/mega-projects-minting-millionaires-in-villages.

Mendes, P.A., Hall, J., Matos, S. and Silvestre, B., 2014. Reforming Brazil's Offshore Oil and Gas Safety Regulatory Framework: Lessons from Norway, the United Kingdom and the United States. *Energy Policy*, 74, pp. 443–453.

Mendonça, R.W. and de Oliveira, L.G., 2013. Local Content Policy in the Brazilian Oil and Gas Sectoral System of Innovation. *Latin American Business Review*, 14(3–4), pp. 271–287.

Warner, M., 2011. *Local Content in Procurement: Creating Local Jobs and Competitive Domestic Industries in Supply Chains.* Sheffield, UK: Greenleaf Publishing.

Mining (Local Content) (Amendments) Regulations, 2019.

Mitchell, P., Santi, E. and Lichtenberg, Z., 2005. May. The Role of Socio-Political Risk Analysis in Developing Communications Strategies for Controversial Projects: The Case of the West Africa Gas Pipeline Project. In *CommNet Forum* (pp. 9–11). Washington, DC: World Bank.

Mkutu, K., Mkutu, T., Marani, M. and Ekitela, A.L., 2019. New Oil Developments in a Remote Area: Environmental Justice and Participation in Turkana, Kenya. *The Journal of Environment & Development.*

Model Contract 2015, Art. 1.3.15.

Mohammed, I., 1985. The Nigerian Enterprises Promotion Decrees (1972 and 1977) and Indigenisation in Nigeria (Doctoral dissertation, University of Warwick).

Monge-González, R., Rivera, L. and Rosales-Tijerino, J., 2010. Productive Development Policies in Costa Rica: Market Failures, Government Failures, and Policy Outcomes.

Moran, T., 2014. Foreign Investment and Supply Chains in Emerging Markets: Recurring Problems and Demonstrated Solutions. Peterson Institute for International Economics Working Paper (2014-12).

Moran, T.H., Graham, E.M. and Blomström, M.(Eds.), 2005. *Does Foreign Direct Investment Promote Development?* Peterson Institute.

Morris, M. and Fessehaie, J., 2014. The Industrialisation Challenge for Africa: Towards a Commodities Based Industrialisation Path. *Journal of African Trade*, 1(1), pp. 25–36.

Morris, M., Kaplinsky, R. and Kaplan, D. (2011a) One Thing Leads to Another Commodities, Linkages and Industrial Development—A Conceptual Overview. MMCP Discussion Paper. Milton Keynes and Cape Town, The Open University and University of Cape Town.

Morris, M., Kaplinsky, R. and Kaplan, D., 2011. Commodities and Linkages: Industrialisation in Sub Saharan Africa. MMCP Discussion Paper, No. 13.

Morris, M., Kaplinsky, R. and Kaplan, D., 2012: One Thing Leads to Another—Commodities, Linkages and Industrial Development—A Conceptual Overview, MMCP Discussion Paper, Milton Keynes and Cape Town, The Open University and University of Cape Town.

Morrissey, O., 2012. FDI in Sub-Saharan Africa: Few Linkages, Fewer Spillovers, European *Journal of Development Research*, 24(1), pp. 26–31.

Mugomba, A.T., 1978. Regional Organisations and African Underdevelopment: The Collapse of the East African Community. *The Journal of Modern African Studies*, 16(2), pp. 261–272.

Murungi, C.N., Introducing Local Content Regulation in Kenya's Economy: A Case Study of Kenya's Energy, Oil and Gas Sector. Institute of Diplomacy and International Studies. University of Nairobi. R50/63843/2010.

Ngoasong, M.Z., 2014. How International Oil and Gas Companies Respond to Local Content Policies in Petroleum-Producing Developing Countries: A Narrative Enquiry. *Energy Policy*, 73, pp. 471–479.

Nigerian Oil and Gas Industry Content Development Act, 2010, s 106 [NOGICD Act].

Nordås, H.K., Vatne, E. and Heum, P., 2003. The Upstream Petroleum Industry and Local Industrial Development: A Comparative Study.

Noreng, Ø., 2004. Norway: Economic Diversification and the Petroleum Industry. *Middle East Economic Survey*, 47(45).

Noreng, Ø., 2009. Brazil and Norway–Offshore Petroleum Experiences and Lessons. The *Journal of Energy and Development*, 35(1/2), pp. 79–99.

NRGI, From Reporting to Reform: Eleven Opportunities for Increasing EITI Impacts, 2015. Available at: https://resourcegovernance.org/analysis-tools/publications/reporting-reform-eleven-opportunities-increasingeiti-impacts.

Ntara, C.K., 2016. African Trading Blocs and Economic Growth: A Critical Review of the Literature. *International Journal of Developing and Emerging Economies*, 4(1), pp. 1–21.

Nwaobi, G.C., 2004. Corruption and Bribery in the Nigerian Economy. Available at SSRN 506342.

Nwapi, C., 2015. Defining the "Local" in Local Content Requirements in the Oil and Gas and Mining Sectors in Developing Countries. *Law and Development Review*, 8(1), pp. 187–216.

Nwapi, C., 2016. A Survey of the Literature on Local Content Policies in the Oil and Gas Industry in East Africa. SPP Research Paper (9/16).

Obiri KA, Bjeirmi B, Boateng, P., 2019. Local Content Implementation Enhancement Through Infrastructure Development in Ghana's Oil and Gas Industry. *Journal of Energy Research and Reviews*, 2019, pp. 1–10.

Ochen, E.A., 2017. The Military in Uganda: War, Peace and Support for Democratic Consolidation. In *African Peace Militaries* (pp. 54–78). Routledge.

ODI Report, May 2014, p. 35, online: http://www.odi.org/sites/odi.org.uk/files/odi-assets/publications-opinion-files/8945.pdf.

Odour, C., et al., 2014. The Political Economy of the Extractive Sector. In Miriam W. Oiro Omolo and Germano Mwabu (Eds.), *A Primer to the Emerging Extractive Sector in Kenya: Resource Bliss, Dilemma or Curse* (p. 181). Nairobi: Institute of Economic Affairs.

Ogbuagu, C.S., 1983. The Nigerian Indigenization Policy: Nationalism or Pragmatism? *African Affairs*, pp. 241–266.

Oloruntoba, S.O., 2019. Innovation and Regional Integration in Africa: Exploring Theory and Praxis for Socio Economic Development. In *Innovation, Regional Integration, and Development in Africa* (pp. 29–41). Cham: Springer.

Ouedraogo, N.S., 2013. Energy Consumption and Economic Growth: Evidence from the Economic Community of West African States (ECOWAS). *Energy Economics*, 36, pp. 637–647.

Outlook, A.E., 2018. African Economic Outlook 2018. Organization for Economic Cooperation and Development OECD, Development Centre, Paris, June.

Ovadia, J.S., 2012. The Dual Nature of Local Content in Angola's Oil and Gas Industry: Development vs. Elite Accumulation. *Journal of Contemporary African Studies*, 30(3), pp. 395–417.

Ovadia, J.S., 2013. The Making of Oil-Backed Local Capitalism in Nigeria. *New Political Economy*, 18(2), pp. 258–283.

Ovadia, J.S., 2014. Local Content and Natural Resource Governance: The Cases of Angola and Nigeria. *The Extractive Industries and Society*, 1(2), pp. 137–146.

Ovadia, J.S., 2016a. Local Content Policies and Petro-Development in Sub-Saharan Africa: A Comparative Analysis. *Resources Policy*, 49, pp. 20–30.

Ovadia, J.S., 2016b. Oil-Backed Capitalist Development in the Global South: A Case of Positive Oil Exceptionalism. In *Energy, Capitalism and World Order: Towards a New Agenda in International Political Economy*. Palgrave Macmillan.

Ovadia, J.S., 2017. Local Content in Tanzania's Gas and Minerals Sectors: Who Regulates? CMI Brief.

Overview of the Brazilian Oil and Gas Industry. http://www.offshorecenter.dk/filer/files/Project/Internationalisering/OCD%20report%20Brazil.pdf.

Oyewole, B., 2018. Overview of Local Content Regulatory Frameworks in Selected ECCAS Countries.

Parshotam, A., 2018. Can the African Continental Free Trade Area Offer a New Beginning for Trade in Africa?

Patey, L., Kenya: An African Oil Upstart in Transition. Oxford Institute for Energy Studies Paper: WPM 53, p. 23, online: http://www.oxfordenergy.org/wpcms/wp-content/uploads/2014/10/WPM-53.pdf.

Paul, I.A., 2015. A Historical Perspective of Petroleum on Nigeria's Economy Crisis since Independence. *Global Journal of Human-Social Science Research*.

Pegg, S., 2006. Can Policy Intervention Beat the Resource Curse? Evidence from the Chad–Cameroon Pipeline Project. *African Affairs*, 105(418), 1–25.

Pegram, J., Falcone, G. and Kolios, A., 2020. Job Role Localisation in the Oil and Gas Industry: A Case Study of Ghana. *The Extractive Industries and Society*, 7(2), pp. 328–336.

Petroleum (Local Content and Local Participation) Regulations, 2013. LI 2204, Reg 49. Available at: http://www.reportingoilandgas.org/wp-content/uploads/PETROLEUMLOCAL-CONTENT-AND-LOCAL-PARTICIPA TION-REGULATIONS2013.pdf.

Phase One of Regional Power Pool Project Set for Completion by Mid-2019. https://www.theeastafrican.co.ke/business/Phase-One-of-regional-power-pool-project-set-for-completion/2560-4699790-g706gj/index.html.

Presidential Regulation No.54/2010 and The Oil and Gas Law No.22/2001 Segregating Regulatory Roles and Commercial Operations.

Raji, A.O.Y. and Abejide, T.S., 2014. The British Mining and Oil Regulations in Colonial Nigeria c. 1914–1960s: An Assessment. *Singaporean Journal of Business, Economics and Management Studies*, 51(1447), pp. 1–14.

Ramdoo, I., 2015a. Resource-Based Industrialisation in Africa Optimising Linkages and Value Chains in the Extractive Sector, Discussion Paper, No. 179, European Centre for Development Policy Management.

Ramdoo, I., 2015b. Synergising and Optimising Mineral Infrastructure in Regional Development Strategies. By International Centre for Trade and Sustainable Development (ICTSD) 7.

Ramdoo, I., 2015c. Unpacking Local Content Requirements in the Extractive Sector: What Implications for the Global Trade and Investment Frameworks. By International Centre for Trade and Sustainable Development (ICTSD).

Ramdoo, I., 2016. Local Content, Trade and Investment: Is There Policy Space Left for Linkages Development in Resource Rich Countries? (No. 205). Discussion Paper.

Ramírez-Cendrero, J.M. and Wirth, E., 2016. Is the Norwegian Model Exportable to Combat Dutch Disease? *Resources Policy*, 48, pp. 85–96.

Ravenhill, F.J., 1974. Military Rule in Uganda: The Politics of Survival. *African Studies Review*, 17(1), pp. 229–226.

Ravenhill, J., 1979. Regional Integration and Development in Africa: Lessons from the East African Community. *Journal of Commonwealth & Comparative Politics*, 17(3), pp. 227–246.

Relaxing of Brazil Local Content Rules Will Boost Oil Output Capacity: Report. https://www.reuters.com/article/us-brazil-presalt/relaxing-of-brazil-local-content-rules-will-boost-oil-output-capacity-report-idUSKBN1JG28T.

Resource-Rich Africa Can Boost Growth to Reduce Poverty. https://www.brookings.edu/blog/future-development/2018/09/25/resource-rich-africa-can-boost-growth-to-reduce-poverty/.

Rocha, F. and Ruiz, A.U., 2011. The Role of the Network Coordinator in the Attraction of Foreign Investment in R&D: The Case of the Brazilian Oil and Gas Industry. *Transnational Corporations*, 20(3), pp. 33–61.

Rodriguez-Clare, A., 2007. Clusters and Comparative Advantage: Implications for Industrial Policy. *Journal of Development Economics*, 82(1), pp. 43–57.

Rodrik, D., 2006. What's So Special About China's Exports? *China & World Economy*, 14(5), pp. 1–19.

Rodrik, D., 2008. Normalizing Industrial Policy. Commission on Growth and Development Working Paper, No. 3. World Bank, Washington, DC. © World Bank. https://openknowledge.worldbank.org/handle/10986/28009. License: CC BY 3.0 IGO.

Royal Decree of 9 April 1965 Relating to Exploration for and Exploitation of Petroleum Deposits in the Sea-Bed and Its Subsoil on the Norwegian Continental Shelf.

Sasson, A. and Blomgren, A., 2011. Knowledge Based Oil and Gas Industry. Research Report 3/2011.

Sauvé, P., 2016. Life Beyond Local Content: Exploring Alternative Measures of Industry Support in the Context of WTO Accession. *Journal of International Trade*, 1, pp. 1–28.

Schlosberg, D., and Carruthers, D., 2010. Local Struggles, Environmental Justice, and Community Capabilities. *Environmental Politics*, 10(4), pp. 12–35.

Schlosberg, D., 2003. The Justice of Environmental Justice: Reconciling Equity, Recognition, and Participation in a Political Movement. *Moral and Political Reasoning in Environmental Practice*, 77, p. 106.

Sen, R., 2018. Enhancing Local Content in Uganda's Oil and Gas Industry (No. 2018/110). WIDER Working Paper.

Stephenson, S., 2013, July 25. Addressing Local Content Requirements: Current Challenges and Future Opportunities. *Biores*, 7(3).

Solutions. Briefing No. 5, Local Content Solutions. http://www.localcontentsolutions.com/pdf/SolutionsSeries5.pdf.

South Sudan Ready to Join East African Community. https://allafrica.com/view/group/main/main/id/00013706.html.

Sovacool, B.K. and Dworkin, M.H., 2015. Energy Justice: Conceptual Insights and Practical Applications. *Applied Energy*, 142, pp. 435–444.

Sovacool, B.K., Heffron, R.J., McCauley, D. and Goldthau, A., 2016. Energy Decisions Reframed as Justice and Ethical Concerns. *Nature Energy*, 1(5), p. 16024.

Sovacool, B.K., Walter, G., Van de Graaf, T. and Andrews, N., 2016. Energy Governance, Transnational Rules, and the Resource Curse: Exploring the Effectiveness of the Extractive Industries Transparency Initiative (EITI). *World Development*, 83, pp. 179–192.

Holden, S., 2011, June. Avoiding the Resource Curse the Case Norway Department of Economics, University of Oslo.

Stiglitz, J.E. and Dasgupta, P., 1981. Resource Depletion Under Technological Uncertainty.

Sturesson, A. and Zobel, T., 2015. The Extractive Industries Transparency Initiative (EITI). In Uganda: Who Will Take the Lead When the Government Falters? *The Extractive Industries and Society*, 2(1), pp. 33–45.

Sullivan, D. and Tifft, L., 2006. Introduction: The Healing Dimension of Restorative Justice: A One-World Body. *Handbook of Restorative Justice*, pp. 1–16.

Taylor, S.D., 2018. Africa and Foreign Direct Investment. In *The Palgrave Handbook of African Politics, Governance and Development* (pp. 709–722). New York: Palgrave Macmillan.

The Brazilian Petroleum Law Number 9478/97.

The East African Community Industrialisation Policy in Brief East African Community Opening new Opportunities for Growth and Expansion of Cross Border Manufacturing and Upgrading of Small and Medium Enterprises (SME). https://eacgermany.org/wp-content/uploads/2014/10/EAC-Industrialization-Policy-Brief.pdf.

The OECD Guidelines and Other Corporate Responsibility Instruments. https://www.oecd.org/corporate/mne/WP-2001_5.pdf.

The Petroleum (Exploration, Development and Production) Act, 2013.

Thurber, M.C., Hults, D.R. and Heller, P.R., 2011. Exporting the "Norwegian Model": The Effect of Administrative Design on Oil Sector Performance. *Energy Policy*, 39(9), pp. 5366–5378.

Todd, H. and Zografos, C., 2005. Justice for the Environment: Developing a Set of Indicators of Environmental Justice for Scotland. *Environmental Values*, 14(4), pp. 483–501.

Tordo, S., 2011. *National Oil Companies and Value Creation.* The World Bank.

Tordo, S. and Anouti, Y., 2013. *Local Content Policies in the Oil and Gas Sector: Case Studies*. World Bank Publications.

Tordo, S., Warner, M., Manzano, O. and Anouti, Y., 2013. *Local Content Policies in the Oil and Gas Sector*. The World Bank.

Uganda and Kenya Local Content Regulations of 2018.

Uganda Local Content Bill, 2017.

Uganda-Tanzania Crude Pipeline Talks Continue. https://www.theeastafrican.co.ke/business/Uganda-Tanzania-crude-pipeline-talks-continue/2560-496 2886-33aijvz/index.html.

UNCTAD, 2010. *Creating Business Linkages: A Policy Perspective*. Geneva: UNCTAD.

UNCTAD, 2013. *Commodities and Development Report: Perennial Problems, New Challenges and Evolving Perspectives*. United Nations Publication.

UNCTAD, World Investment Report 2015: Reforming International Investment Governance. http://unctad.org/en/PublicationsLibrary/wir2015_overview_en.pdf. That is presuming that local companies have the competency to meet the requirements for the good or service in the first place.

Union, A., 2019. Protocol on Relations Between the African Union (au) and the Regional Economic Communities (RECs).

United Republic of Tanzania, Local Content Policy, 2014.

Valerie Marcel and Others, A Local Content Decision Tree for Emerging Producers (Energy, Environment and Resources Department Chatham House: The Royal Institute of International Affairs 2016).

Victor, N.M., 2007. On Measuring the Performance of National Oil Companies (NOCs). Program on Energy and Sustainable Development Working Papers, WP64, Stanford University.

Votaw, D., 1972. Genius Becomes Rare: A Comment on the Doctrine of Social Responsibility Pt. I. *California Management Review*, 15(2), pp. 25–31.

Walker, G., 2009. Beyond Distribution and Proximity: Exploring the Multiple Spatialities of Environmental Justice. *Antipode*, 41(4), pp. 614–636.

Walker, G., 2012. *Environmental Justice: Concepts, Evidence and Politics*. Routledge.

Warner, M., 2007. Community Content: The Interface of Community Investment Programmes with Local Content Practices in the Oil and Gas Development Sector. Overseas Development Institute (ODI) Briefing Note, (9).

Warner, M., 2010. Unpacking Local Content Metrics and Measurement Dr Michael Warner, November PDF. [online] Docplayer.net. Available at: https://docplayer.net/48696189-Unpacking-local-content-metrics-and-measurement-dr-michael-warner-november-2010.html.

Warner, M., 2011. *Local Content Solutions: Participation of Domestic Industry in Procurement for Oil, Gas and Mining Projects*. Sheffield: Greenleaf Publishing.

Warner, M., 2017. *Local Content in Procurement: Creating Local Jobs and Competitive Domestic Industries in Supply Chains*. Routledge.

Wellisz, S. and Findlay, R., 1988. The State and the Invisible Hand. *The World Bank Research Observer*, 3(1), pp. 59–80.

Zacour, C., Zuma Pereira, T., Lima Rocha Cristofaro, A. and Ferreira Francisco, F., 2012. Petrobras and the New Regulatory Framework for the Exploration and Production of Oil and Natural Gas in the Brazilian Pre-salt Region. *Journal of World Energy Law and Business*, 5(2), pp. 125–138.

Index

R. S. Muhongo, *Energy Justice*, Energy, Climate and the Environment,
https://doi.org/10.1007/978-3-030-61338-9

The manufacturer's authorised representative in the EU is Springer Nature Customer Service Centre GmbH, Europaplatz 3, 69115 Heidelberg, Germany. If you have any concerns regarding our products, please contact ProductSafety@springernature.com

Printed and bound by CPI Group (UK) Ltd, Croydon, CR0 4YY
23/09/2026
02221544-0004